First in Line

First in Line

PRESIDENTS, VICE PRESIDENTS, AND THE PURSUIT OF POWER

KATE ANDERSEN BROWER

HARPER

An Imprint of HarperCollins*Publishers*

HarperCollins books may be purchased for educational, business, or sales promotional use. For information, please email the Special Markets Department at SPsales@harpercollins.com.

FIRST EDITION

Library of Congress Cataloging-in-Publication Data has been applied for.

ISBN 978-0-06-266894-3

18 19 20 21 22   LSC   10 9 8 7 6 5 4 3 2 1

For my radiant sister Kelly

&

For those who have patiently served:
Walter Mondale, George H. W. Bush, Dan Quayle, Al Gore,
Dick Cheney, Joe Biden, and Mike Pence

CONTENTS

PRESIDENTIAL PARTNERS

RICHARD M. NIXON AND DWIGHT D. EISENHOWER,	1953–1961
LYNDON B. JOHNSON AND JOHN F. KENNEDY,	1961–1963
HUBERT H. HUMPHREY AND LYNDON B. JOHNSON,	1965–1969
SPIRO T. AGNEW AND RICHARD M. NIXON,	1969–1973
GERALD R. FORD AND RICHARD M. NIXON,	1973–1974
NELSON ROCKEFELLER AND GERALD R. FORD,	1974–1977
WALTER MONDALE AND JIMMY CARTER,	1977–1981
GEORGE H. W. BUSH AND RONALD REAGAN,	1981–1989
DAN QUAYLE AND GEORGE H. W. BUSH,	1989–1993
AL GORE AND BILL CLINTON,	1993–2001
DICK CHENEY AND GEORGE W. BUSH,	2001–2009
JOE BIDEN AND BARACK OBAMA,	2009–2017
MIKE PENCE AND DONALD TRUMP,	2017–

ON THE VICE PRESIDENCY

—————————————

I do not propose to be buried until I am really dead and in my coffin.

—MASSACHUSETTS SENATOR DANIEL WEBSTER'S REPLY WHEN OFFERED
THE RUNNING-MATE SPOT ON THE WHIG PARTY TICKET IN 1848

I would a great deal rather be anything, say professor of history, than Vice-President.

—THEODORE ROOSEVELT, VICE PRESIDENT
UNDER PRESIDENT WILLIAM MCKINLEY

Once there were two brothers: one ran away to sea, the other was elected vice president. Nothing was ever heard from either of them again.

—THOMAS R. MARSHALL, VICE PRESIDENT
UNDER PRESIDENT WOODROW WILSON

The vice presidency is not worth a bucket of warm piss [later cleaned up to "warm spit"].

—JOHN NANCE GARNER, VICE PRESIDENT
UNDER PRESIDENT FRANKLIN D. ROOSEVELT

The vice presidency is "that rare opportunity in politics for a man to move from a potential unknown to an actual unknown."

—SPIRO AGNEW, VICE PRESIDENT
UNDER PRESIDENT RICHARD NIXON

Dick, I don't know how you ever took the job.

—GERALD FORD, WHO SPENT LESS THAN A YEAR AS
RICHARD NIXON'S VICE PRESIDENT, TO DICK CHENEY,
VICE PRESIDENT UNDER PRESIDENT GEORGE W. BUSH

You die, I fly.

—GEORGE H. W. BUSH, VICE PRESIDENT UNDER PRESIDENT RONALD
REAGAN, ON VICE PRESIDENTS BEING DISPATCHED AROUND THE WORLD
TO ATTEND DIGNITARIES' FUNERALS

*Let me make this pledge to you right here and now: For every American
who is trying to do the right thing, for all those people in government who
are honoring their pledge to uphold the law and respect our Constitution, no
longer will the eight most dreaded words in the English language be: "The
vice president's office is on the phone."*

—JOE BIDEN, VICE PRESIDENT UNDER PRESIDENT BARACK OBAMA,
LAMPOONING HIS PREDECESSOR, DICK CHENEY

Iron sharpens iron.

—MIKE PENCE, VICE PRESIDENT UNDER PRESIDENT DONALD TRUMP,
QUOTING FROM HIS FAVORITE BIBLE VERSE, PROVERBS 27:17:
"AS IRON SHARPENS IRON, SO ONE MAN SHARPENS ANOTHER."

"Sue, did the president call?
"No."

—SELINA MEYER, FICTIONAL ONETIME VICE PRESIDENT
ON HBO'S *VEEP,* TO HER BELEAGUERED SECRETARY, SUE

First in Line

They felt called to do this. They knew what they were
getting into, they knew the history, they knew the challenges,
and they accepted that as part of their calling.

—JIM ATTERHOLT, MIKE PENCE'S GUBERNATORIAL CHIEF OF STAFF, ON
MIKE AND KAREN PENCE'S DECISION TO JOIN DONALD TRUMP'S TICKET

The Trumps were hunkered down at their golf club in Bed-minster, New Jersey, in the summer of 2016. It was the final meeting in a series of discussions to decide on Donald Trump's running mate, and, as always, it was a family affair. Some combination of Trump's eldest children—Donald Jr., Eric, Ivanka—and Ivanka's husband, Jared Kushner, had been mainstays at meetings with Washington lawyers in charge of vetting vice presidential candidates. But at this final decisive meeting it was Melania Trump, the aloof former model married to the outspoken and impulsive real estate tycoon, who drew the bottom line. Whoever is chosen must be "clean," she insisted. That meant no affairs and no messy financial entanglements. In short, it meant no drama. She realized that her husband had a surplus of that already.

Melania was an important voice in the room during that last critical meeting, even though she was conspicuously absent when her husband actually announced Indiana governor Mike Pence as his running mate. It was the first time in modern campaign history that the wife of a presidential candidate was not at the public announcement, and it was an early indication of how uncomfortable she would be as first lady. It was decided at that final meeting that what they needed was someone with "safe hands," as vetting lawyers call it. Someone who would be calm in a crisis; someone who could instill a sense of confidence in the Republican base that remained deeply skeptical of Trump. Most of all, what they needed was someone who could take over the presidency, if necessary.

Melania was keenly aware of the need to balance her husband, who has spent much of his public life—and most of his life was lived clinging to the spotlight—awash in scandal. She wanted to make sure that there were absolutely no skeletons in his running mate's closet. But one finalist had a closet full of them (still, Donald Jr. backed him until the end), and another contender was so controversial that he would be ousted within the first few weeks of the administration when he served in a different position. Melania's shrewd instincts proved correct; Mike Pence was by far the least controversial on Trump's list of vice presidential candidates, and Pence could help Trump win over conservative Republicans. Melania is described by people who know her as "stubborn" and "unapologetic about who she is." "No one speaks for me," Melania once said when her husband promised a TV news anchor that she would do her show. In this case, she was decidedly in Pence's corner.

Trump came late to the search for a running mate and did not reach out to Arthur B. Culvahouse, the well-connected Republican lawyer who led the vetting for John McCain in 2008, until late May. Trump's campaign chair, Paul Manafort, even considered paying a law firm to do the vetting, seemingly unaware of the long-held tradition of lawyers in Washington and New York clamoring to do it for free. Lawyers put together detailed reports on each of the candidates, including their tax returns and any history of psychiatric treatment, and they dig into rumors of affairs. In that secretive vetting ritual, Culvahouse makes a point of only using lawyers from his own Washington firm to guard against leaks. Kushner, the then-thirty-five-year-old real estate scion married to Ivanka, teased Culvahouse that one of his write-ups on a candidate read "like a legal treatise," and another "like the script for *House of Cards*."

Trump's options were limited. "Trump was hard to get your mind around if you're vetting vice presidential candidates because he had made a number of provocative statements that would be potentially disqualifying in a conventional vice presidential nominee," said Culvahouse, who was White House counsel to Ronald Reagan and contributed to Jeb Bush's and Marco Rubio's 2016 campaigns. A couple of Trump's picks, including Republican senator Bob Corker of Tennessee, took themselves out of the running, not because of personal entanglements, but because of moral objections—they felt they could not defend Trump every day, which is a key element of the vice presidency. The two finalists who would be one heartbeat away from the presidency could not have been more different, both in temperament and reputation.

Trump crowdsourced the process, asking anyone and everyone he met who he should pick. And even though he never released his own tax returns, Trump asked for his potential running mate's financial information. He was looking for someone who fit the part, someone who *looked* like a vice president. "Straight from central casting," Trump is reported to have said of Pence. Culvahouse said Trump's long list of candidates was much shorter than McCain's (nominees have a longer list of names at the beginning of their search and a whittled down, shorter list toward the end of the process). McCain had almost twenty-five people on his long list, and Trump had just ten on his, including Michael Flynn, a controversial retired U.S. Army lieutenant general and former intelligence officer who was once the director of the Defense Intelligence Agency during Barack Obama's administration.

"He [Trump] was clearly fond of Flynn," Culvahouse said, shaking his head. Even though Culvahouse says he did not interview Flynn for the position, Flynn remained on Trump's list for a while, no matter how many people tried to talk Trump out of it. There was some discussion among Trump's campaign staff and Culvahouse's team of lawyers about whether Flynn was actually a registered Democrat until someone on Culvahouse's staff produced a photo of his voter ID card confirming it. It turns out, of course, that that was the least of Flynn's problems. Though Flynn did not make it on Trump's short list of VP candidates, Trump made him national security adviser, where he served for less than a month before being forced out. Flynn is one of the key figures in special counsel Robert Mueller's investigation into Russia's interference in the 2016 election and the Trump campaign's alleged ties to Russian meddling. By the end of 2017, Flynn had pled guilty to making false statements to the FBI about his conversations during the transition period with Russia's ambassador to the United States.

Unlike with Cabinet picks, the FBI does not do a background check on vice presidents—when Trump picked Rex Tillerson as secretary of state, the FBI did a background check, but not so with Trump's number two, the man who is next in line for the presidency. "The real problem with vice presidential vetting that just terrifies me, Flynn is the best example, is that you don't have the FBI to help you," Culvahouse said, exasperated. Unlike a Cabinet officer, whom a president can fire, there is no way for a president to easily remove his vice president from office—only Congress can impeach him. The FBI is not involved in the vice presidential vetting process because nominees for vice president (like nominees for president) are not technically being considered for federal office, they're being considered for the nomination of a political party. "It's a huge hole I think in how we pick our vice presidents," Culvahouse said. "It absolutely should be changed." The FBI has considerably more resources than law firms do, and in the case of Flynn, likely would have picked up troubling signs. Armed with that information the FBI could have flagged the campaign to go slow and think twice before giving him a role in the administration.

In the end, it was down to two men. Mike Pence, a devout evangelical Christian in his late fifties, won out over former speaker of the House Newt Gingrich, the much more controversial finalist, who, like Trump, was also in his seventies and had been married three times. Gingrich left his first wife when she was in the hospital recovering from cancer surgery and did not take his seat for a third term as Speaker in part because of ethics violations. He was far from "clean." Pence had served six terms in the House, had strong ties to Republican leaders, and, most important, he could help Trump win votes in the Midwest.

David McIntosh, a friend of Pence's, is a former Indiana congressman who now heads the influential conservative group the Club for Growth: "Trump needs Pence there as a less mercurial and more stable conservative leader," he said. Trump, who had very little knowledge of how Capitol Hill works, told Culvahouse he wanted Pence to be the COO, or chief operating officer, of the White House. In this redefinition of the executive branch, Trump, then, would be the CEO of the United States—an unprecedented approach to the presidency. But how would someone so different from the man he was being asked to serve respond to the offer? One longtime friend of Pence's said that Pence considered the vice presidency his "divine appointment." Pence told another close friend, "It isn't about Donald Trump. It's about the country."

Two years before Pence became vice president he and his friend then–Indiana senator Dan Coats talked privately about their political futures. Pence was weighing whether to run for governor for a second term or to seek the presidency in 2016 and Coats was trying to decide whether to run for reelection to the Senate. "We talked about the future and where God might lead each of us," Coats recalled. "We prayed that God would be clear, and I think I raised the question that we should pray for clarity, not for what *we* want but clarity for what *God* would want. I'm always a little hesitant to discuss it in these terms because people say, 'Oh, you think you were ordained.' That's not it at all, I think we both feel that it was a question of how God could best use our talents in whatever direction He wants to take us . . . a whole number of miraculous things happened in the political world that affected both of our lives."

Coats went on to reluctantly accept Trump's offer to head the intelligence community as director of national intelligence only after Pence repeatedly asked him to take the job. "We need someone with experience," Pence pleaded with him. Both these men see themselves as bulwarks against chaos.

The "miraculous" offer to be Trump's running mate would rock the Pences' lives. Pence likes to say, "There are two types of people in the world: Those who are called and those who are driven." But it is always a little bit of both. Jim Atterholt, who was Pence's chief of staff when he was governor of Indiana, said Pence and his wife, Karen, "prayed a great deal" when they were considering the job. Karen is her husband's top adviser and the Pences are a "buy one, get one free package," not far off from Bill and Hillary Clinton. Pence said he had two conditions before accepting Trump's offer: Their families had to get to know each other and he needed to understand the job description. Trump replied, "I'm going to have the most consequential vice president ever. That's what I want. That's the job description." It was music to Pence's ears. Pence called Trump with Karen on the line to accept his offer. There was no specific agreement reached between Pence and Trump about what Pence's role would be as vice president, but in the end there was very little hesitation on the part of both Pences. "They felt called to do this," Atterholt recalled. "They knew what they were getting into, they knew the history, they knew the challenges, and they accepted that as part of their calling."

But Pence was not truly expecting to win—no one was. A week after Trump's shocking victory, Trump was still asking friends, *Can you believe I won this thing?* Before election day, one of Hillary Clinton's campaign aides approached the residence manager at the Naval Observatory, where the vice president lives, to see if he would agree to stay on and help transition Clinton's running mate, Virginia senator Tim Kaine, and his family into the residence. No one from the Trump campaign ever bothered to call.

I

You're a Guest in My House

I am Vice President. In this I am nothing, but I may be everything.

—JOHN ADAMS, AMERICA'S FIRST VICE PRESIDENT, WHO
SERVED UNDER PRESIDENT GEORGE WASHINGTON

The vice presidency plays head games with you. You
start to wonder if you matter that much.

—MIKE DONILON, A LONGTIME CONFIDANT OF JOE BIDEN'S

Even the best relationships between modern presidents and their vice presidents can be difficult at times, especially when they were once rivals for the nomination. Barack Obama roundly beat Joe Biden in the 2008 Democratic primaries and aides say Biden always respected Obama because of that. "Joe Biden is not a humble person in some ways and he has great respect for his own judgment," quipped Ron Klain, who was chief of staff to Biden and Al Gore, "but he does believe in the democratic process, so there was never any confusion in his mind about how all this worked out." Obama was a global superstar who had run laps around Delaware's longest-serving senator. "I don't think on his best day Joe Biden believes that Barack Obama won in 2008 because of Joe Biden," Klain added. So, he said, Obama clearly had the upper hand in their relationship. According to Klain, Obama's perspective was simple, if not a bit condescending: *This is my house, these are my things, I'm interested in your views, Joe, I like your input, I want you to be happy here, but you're a guest in my house.*

The thirteen men in this book, from Richard Nixon to Mike Pence, eight Republicans and five Democrats, all share one thing in common: the deeply humbling experience of being the president's understudy. When asked for his greatest accomplishment during his eight years as Ronald Reagan's vice president, George H. W. Bush, still not wanting to take credit for any achievements, replied: "That is for others to judge." Vice presidents have only two constitutional duties: to succeed the president if he is unable to serve for any reason, and to act as president of the Senate to cast tiebreaking votes. (Biden cast none, whereas Mike Pence cast nine tiebreaking votes in just over a year on the job.) No matter how integrated they are, vice presidents always play second fiddle; at best they are the politician who was *almost* good enough, at worst they are a virtual unknown who embarrasses the president.

The vice presidency has been a much-maligned and oft-deprecated position. An important member of the constitutional convention, Roger Sherman, thought the vice president would have to be assigned the role of presiding officer of the Senate because, he said, without it the vice president "would be without employment." Benjamin Franklin suggested that the vice president be addressed as "Your Superfluous Excellency." Warren Harding's vice president, Calvin Coolidge, said: "I enjoyed my time as vice president. It never interfered with my mandatory eleven hours of sleep a day." It is indeed an often-thankless job, the butt of countless jokes, many of which are from vice presidents themselves, from both parties. Harry Truman, who served for less than three months as President Franklin D. Roosevelt's vice president, complained, "The vice president simply presides over the Senate and sits around hoping for a funeral." Upon hearing the news of a foreign leader's death George H. W. Bush would tell aides, only half-joking, "You die, I fly." When President Theodore Roosevelt grew irritated by the tinkling of a glittering eleven-foot-high chandelier, purchased by Thomas Jefferson and hanging in the Oval Office, Roosevelt said, "Send it up to the Vice President. He never has anything to do and the tinkling will keep him awake!" The chandelier was brought to the vice president's Senate office (the vice president at the time was Charles W. Fairbanks) and remained there until then–vice president Lyndon B. Johnson got rid of it.

The job occupies what the late journalist David S. Broder called "the strange no-man's-land between legislative and executive power." It comes with a job description that is only slightly more defined than that of first lady—which has none whatsoever. Historian Arthur Schlesinger Jr. called the vice presidency "a job of spectacular and, I believe, incurable frustration." In a vivid example of the office's muddled position, the vice president as president of the Senate is paid by the Senate and makes $230,700 a year compared to the president's $400,000 (for Mike Pence, $230,700 is more than twice what he made as governor). His retirement income is paid by the Senate as well, but, unlike the president's, it is not guaranteed. In order to receive retirement benefits, the vice president would either need to have served in Congress or in some federal office for at least five years, or served more than one term as vice president. When they leave office, vice presidents are given a budget to help them ease back into civilian life, but even that budget is allocated to them by the presidents they served. When they left office, President George H. W. Bush and Vice President Dan Quayle were awarded a total of $1.5 million, of which Bush gave Quayle and his staff $250,000. And unlike former presidents, who receive lifetime Secret Service protection for themselves and their spouses should they want it after they leave office (the children of former presidents are allowed protection until they are sixteen years old), vice presidents and their spouses and children under sixteen years of age only get Secret Service protection for up to six months after they leave the White House.

But in recent decades the vice presidency has been nothing to laugh at. Like the president, modern vice presidents are followed twenty-four hours a day by a military aide who carries a forty-five-pound aluminum briefcase (called the "football") and they carry a card ("the biscuit") that contains the nuclear verification codes used to authorize a nuclear strike. If the president were to become incapacitated, the vice president would have access to the "football" and the authority to order a launch. In a 2008 interview, Dick Cheney said of that power: "He [the president] could launch a kind of devastating attack the world's never seen. He doesn't have to check with anybody. He doesn't have to call the Congress. He doesn't have to check with the courts. He has that authority because of the nature of the world we live in." The president or vice president, if the president is unable, has an estimated twelve minutes to decide to launch missiles if the United States is under attack. A nuclear launch now would be unimaginably destructive and much more powerful than the bombs that destroyed the Japanese cities of Hiroshima and Nagasaki in 1945. If the president is dead, or incapacitated in any way, the vice president carries that enormous responsibility.

———

Modern presidents have at least five times the staff of their vice presidents in the White House complex alone, and the rest of the government, including the vice president's staff, technically works for the president. Vice presidents typically have about eighty aides who work out of the Eisenhower Executive Office Building, an enormous granite, slate, and cast-iron post–Civil War–era building next door to the White House. The center of power, though, is the Oval Office, and the most influential vice presidents are the men who have spent the most time there advising the president. They are constantly fighting for "unstaffed" access to the president—or alone time—and watching to make sure they are not excluded from Oval Office and Cabinet meetings. Most vice presidents save disagreements with the president for private conversations. "All of us as human beings would rather have something said in private,"

Al Gore remarked, when asked whether he disagreed with President Clinton in front of others. But no matter how influential, vice presidents are utility players who go to events that do not merit the president's attendance, and when they campaign they often play to big crowds in small ponds. What tends to go wrong in these relationships between the president and the vice president is that this gap eats at the understudy, who almost always sees himself as better equipped for the top job than the president he serves. Biden sometimes got tired of the role, but it rarely affected his relationship with the president because he was so personally fond of Obama. Al Gore, however, had a more difficult time of it.

During the 1992 presidential campaign, Gore's old Harvard professor, the noted political scientist Richard E. Neustadt, sent him cautionary notes about the complexity of the role he was seeking. "The VP reminds the P [president] of his mortality," Neustadt warned him. "The White House staff lives in the present, the VP's staff in the future." Three decades earlier, when it was suggested that Kennedy's vice president, Lyndon Johnson, move from his house, which was miles from downtown Washington, and into a townhouse across the street from the White House, Kennedy moaned, *"You think I want Lyndon [Johnson] listening across the park for my heartbeat? NO!"*

In the modern era, Dick Cheney is the only vice president who had no presidential ambitions because he had abandoned them years earlier. Vice presidents and their advisers insist their primary job is to advance the president's agenda, and not their own. But always, in the back of their minds, the vice president is considering how the president's policies will affect him and his own chances of one day being elected president. "I can't say I was never thinking about my own political career," Jimmy Carter's vice president, Walter Mondale, confessed, "but I believe I made an honest commitment that I would put his presidency above all other considerations and I believe I did do that."

Access and information are key. "The vice president's real power derives from his relationship with the president," says Bruce Reed, who was a policy adviser in the Clinton White House and Biden's chief of staff after Klain left. "In the end, the leash is only as long as the president chooses to make it." Powerful vice presidents, like Gore, Biden, and Cheney, had an agreement with the presidents they served that they would see every piece of paper that came across the president's desk. Biden and Cheney tried not to take on many specific assignments like Gore's mission to "reinvent" government and to cut down on waste; instead, they saw themselves as the president's most influential adviser. Even so, all of these men, at times, felt underutilized. Dan Quayle, who became the butt of endless jokes as VP and was less powerful than most of his fellow vice presidents, said he had more interaction with George H. W. Bush on national security matters when he was a senator on the armed services committee and Bush was VP, than when he was serving as Bush's own vice president.

Even travel for vice presidents is fraught. There is a fleet of planes called the Presidential Airlift Group docked at Joint Base Andrews (formerly Andrews Air Force Base) in Prince George's County, Maryland, just outside of Washington. They work almost like a carpool for VIPs and are made available to the vice president, the secretary of state, the first lady, and, on certain occasions, the second lady. Two Boeing VC-25As, referred to as Air Force One when the president is on board, are everyone's first choice. There is a precedent and an order to who gets which plane, and in the past, if then–secretary of state Hillary Clinton had a long trip and Biden was traveling domestically, Biden would get a smaller plane and he would sometimes not be happy about it.

But the vice presidency has its own perks and traditions. Like former presidents, there is a unique camaraderie among vice presidents forged not in the understanding of what it's like to be the most powerful political leader in the world but in the shared experience of proximity to power coupled with the ultimate powerlessness that so often accompanies the vice presidency. Former vice presidents do talk with Mike Pence, but they are reluctant to go into details. One of them explained: "If I tell you and it's written then I'd drastically undercut him. He'd get his head chopped off [by Trump]," he said, swiping his hand across his neck. They see him as a stabilizing force in an unstable White House. Cheney has been advising Pence, who finds himself in an incredibly perilous position, not wanting to lose influence by upsetting Trump while also not wanting to stand by statements from his boss that might hurt his own chances at the presidency. Cheney was not as interested in advice when he was vice president. A couple of months after taking office, he invited Richard Moe, who was Vice President Walter Mondale's chief of staff, for lunch at the White House to go over Mondale's approach to the vice presidency. Moe remembers a lot of long silences. "My lasting impression of that was that Cheney was very incurious," he said. "I don't know why he was going through the motions. He clearly had his own idea of what to do. He had his own power center and a separate policy center." Cheney's chief of staff Scooter Libby did most of the talking.

Pence has been seeking advice from one very unlikely former vice president: Joe Biden. Biden has become one of Donald Trump's most vocal critics. Shortly after the deadly 2017 white nationalist protest in Charlottesville, Virginia, Biden wrote in an *Atlantic* op-ed, "We are living through a battle for the soul of this nation." But Biden and Pence talk at least once a month, which stands in stark contrast to Trump and his predecessor, Barack Obama, who have not spoken since the inauguration, nor have their wives, Melania Trump and Michelle Obama. Biden says Pence told him how much he admired his relationship with Obama and asked for his advice on the office. Pence calls him occasionally to ask questions about foreign policy. "Joe," he says, "this is the decision that has to be made. What went before it?" Biden says he has made himself "available to him on mostly background things, to give him perspective."

Biden has long-standing friendships with foreign leaders from his years in the Senate and as vice president, and he frequently fields calls from worried heads of state. He says he is in touch with a dozen world leaders who ask for advice on how to work with the Trump administration and how to interpret new policies. "The king of Jordan comes to meet with him [President Trump] and gets in a helicopter to come meet with me in Delaware," Biden said, sounding pleased.

During one trip to speak at a conference in Athens, Greece, five months after leaving office, Biden met with Greek prime minister Alexis Tsipras. "I thought it was going to be a ten-minute conversation, and it was two and a half hours," he recalled. "What I do is, whenever that occurs, I pick up the phone and I call Mike. What I tell these leaders, who are somewhat *diffident* about the president, I suggest they go see Mike. I explain it this way: I say that a lot of this doesn't fall within the president's bandwidth, because he has no experience in this area."

Biden is quick to point out how different he and Pence are in their politics, but, like so many others, he treats Pence like a lifeline to a White House in a constant state of upheaval. "Mike and I have talked about it a few times. I have fundamental disagreements with Mike, particularly on social policy and what I consider basic civil rights, civil liberties, but Mike's a guy you can talk with, you can deal with, in a traditional sense. Like [Bill] Clinton could talk to [Newt] Gingrich." Biden says he tries not to criticize the president during private meetings with foreign leaders, though it is not always easy biting his tongue, and in some cases it sounds like he is working as a conduit for the administration. Before he met with former Ukrainian president Viktor Yushchenko, Biden checked in with Pence to make sure he did not misrepresent administration policy and called him after the meeting and gave him a readout.

––––––––

Unlike presidents, who leave notes behind for their successors, vice presidents sign a drawer in their desk in the ceremonial office in the Eisenhower Executive Office Building. The desk was first used by Teddy Roosevelt in 1902 and has been signed by the occupants of the office since the 1940s. And marble busts of the vice presidents are placed throughout the Senate wing of the Capitol. "Being cast in marble is something every vice president looks forward to," said Cheney. "It's not only a high honor, it's our one shot at being remembered."

The vice president is the only member of the president's team who cannot easily be fired (a fact Joe Biden liked to repeat), and he is also usually one of the last people left standing in the administration as other aides either resign out of exhaustion or are pushed out. In the Obama White House it was top adviser Valerie Jarrett and Joe Biden who were among the only people left after eight years. And the relationship between a president and his vice president has historically been forged in fire. The White House is a relatively small place with fewer than a hundred people who are most intimately involved in presidential and vice presidential decision-making. These are weighty decisions and occasionally vice presidents can help share the president's burden.

In nearly every modern administration, beginning with Dwight D. Eisenhower and Richard Nixon, the relationship between the president and his vice president has deteriorated over time. Cheney admits his influence waned in Bush's second term, and the relationship between Gore and Clinton almost fully unraveled at the end of their administration. "I think Vice President Gore was increasingly living in the future, President Clinton was increasingly living in the ticking clock that's created by the Twenty-second Amendment," said Klain, referring to the constitutional amendment that created term limits. Joe Biden and Barack Obama are the only exception to the rule (George H. W. Bush and Ronald Reagan grew closer, but they did not develop a close personal friendship). But while Biden and Obama's relationship did not start out as well as people might think, it grew significantly over their eight years together. "Their friendship is more interesting than it's often portrayed because it was not always easy," said former White House press secretary Josh Earnest. "They were a political odd couple whose strengths were complementary and mutually reinforcing. But there were significant differences in style—and occasionally in perspective." A close aide to Biden said the message was clear from the president's staff from the beginning: *We matter, you don't.* Biden's style clashed with the "no-drama Obama" team. "Like most vice presidential nominations it was kind of a shotgun wedding," said Obama senior campaign and White House aide David Axelrod. "They didn't really know each other well and all of a sudden they were married. That takes some adjustment."

"You know what surprises me," Obama told Biden at one of their weekly lunches, about eight months into the administration.

"No," Biden replied.

"We've really become close friends," Obama said.

"Surprised *you*?" Biden replied with laughter. "What do you mean it surprised *you*?"

They are two very different men. Working for Obama, an aide said, there were days with twelve meetings on twelve different subjects and the ball would be moved forward on each topic of discussion. "When I worked for Biden," the former staffer said, "there were days when I'd spend eight or ten hours just sitting in his office and he'd say, 'Let's have a meeting,' and the meeting would be three hours long." Despite their different management styles, Obama and Biden became remarkably close.

———

Fourteen vice presidents have become president, eight of them ascending to the highest office because of the death of the sitting president. The eight vice presidents who succeeded presidents who died in office are John Tyler (upon William Henry Harrison's death in 1841), Millard Fillmore (upon Zachary Taylor's death in 1850), Andrew Johnson (upon Abraham Lincoln's assassination in 1865), Chester A. Arthur (upon James Garfield's assassination in 1881), Theodore

Roosevelt (upon William McKinley's assassination in 1901), Calvin Coolidge (upon Warren Harding's death in 1923), Harry Truman (upon Franklin Roosevelt's death in 1945), and Lyndon Johnson (upon John F. Kennedy's assassination in 1963).

In the post–World War II era, the vice presidency has become more and more consequential. "Vice presidents are generally an uninteresting lot," Cheney admitted. "There are fascinating relationships now. I think you'd have to say that the really consequential vice presidents are the ones who get to be president"—an ironic statement coming from the most powerful vice president in modern history. Beginning with Harry Truman in 1945 and up until George H. W. Bush, five out of nine presidents were former vice presidents: Truman, Johnson, Nixon, Ford, and Bush—two by election, two by death, and one because of resignation.

In that same time span, three vice presidents became losing nominees for the presidency: Nixon, who lost in 1960; Mondale, who ran and lost four years after serving as Jimmy Carter's VP; and Gore, who lost in 2000. But vice presidents have a huge advantage over their rivals for the highest office if the president they serve allows them to use it: they can do early fund-raising and capitalize on the president's supporters. Yet they have to use their position wisely so that it does not look as if they are preoccupied with their own ambition—either to the president they serve or to the people whose votes they want to receive. It is a difficult needle to thread; only four men have successfully run for president as sitting vice presidents: John Adams, Thomas Jefferson, Martin Van Buren, and Ronald Reagan's vice president, George H. W. Bush. (Richard Nixon ran and won in 1968, eight years after he served as vice president.) "It's been a long time, Marty," Bush said at his first news conference after the 1988 election, when he became the first incumbent vice president to win the presidency since Van Buren in 1836. Part of Bush's victory can be attributed to Ronald Reagan's popularity, and his win was a virtual third term for Reagan. Bush won 80 percent of the votes of those who approved of Reagan's job performance, and lost nine to one among those who disapproved. So while winning their party's nomination is all but guaranteed, vice presidents sometimes have a difficult time closing the deal.

———————

When the founding fathers gathered in Philadelphia in 1787 to write the Constitution, the vice presidency was not at the center of their attention. Article I, Section 3 of the Constitution states that the vice president will preside over the Senate "but shall have no vote, unless they be equally divided." The Founding Fathers, wary of anything resembling a monarchy, were concerned that a president may become too powerful if the vice president had any role in the Senate greater than serving as tiebreaker, so the vice president's job description was kept deliberately diminutive. The line between the executive and legislative branches of government was not to be blurred, and too much power was not to be concentrated in the White House. So, while a vice president can sit in the presiding officer's chair, he cannot speak on the Senate floor without permission. Article II, Section 1 states that the vice president shall succeed the president upon the death, resignation, or removal of the president. Alexander Hamilton of New York wanted the most fundamental aspect of the job—presidential succession—to be strictly temporary. The vice president, Hamilton wrote in a draft of the Constitution, would "exercise all the powers of this Constitution vested in the President until another shall be appointed." The Constitution's wording on succession was never clear until the Twenty-fifth Amendment, which was ratified in 1967, and even then at least one vice president, Dick Cheney, thought it should be made more transparent. George H. W. Bush agrees. Bush has the distinction of serving as president and vice president and says that the Twenty-fifth Amendment should "probably" be revisited so that succession is made more clear.

There was debate among the founders over an even more basic concept: who would elect the president. In their quest to create a democracy and their attempts, however vain, to avoid the creation of competing political parties, some delegates to the Constitutional Convention wanted the members of the national legislature to vote for president, while others argued that it would lead to tyranny if the president was beholden to the members of Congress who had voted for him. So instead they created a system—the electoral college—in which each state has a number of electors equal to the number of the state's senators and members of the House of Representatives. But there was no differentiation between presidential and vice presidential candidates on the ballot, and each elector cast two votes for president (for different candidates, sometimes from different parties) and no vote for vice president. The candidate who won the majority of electoral votes would become president and the runner-up would be named vice president. The House of Representatives would break a tie. As runner-up, the vice president was heir apparent to the presidency, which explains why John Adams and Thomas Jefferson were the nation's first two vice presidents and its second and third presidents. But the system was complicated and led to a president and vice president from two different political parties being elected. In 1796, Adams, a Federalist, was elected president and Jefferson, a Democratic-Republican, was the runner-up and therefore became Adams's vice president. Adams never consulted Jefferson on important national issues and Jefferson used his time as vice president to secure his place as leader of his party. The system was untenable. The presidential election of 1800 was another bitter contest that in the end changed the way a vice president is elected. Instead of becoming vice president as a result of the presidential vote, the next vice president would assume office as part of a party slate. In 1804 the Twelfth Amendment was adopted recognizing the reality of political parties and dictating that each elector cast separate votes for president and vice president. This significantly weakened the position, making the vice president a stand-in and not the second-most-qualified politician for the presidency.

In 1940, Franklin Delano Roosevelt only agreed to seek a third term as president if he could pick Henry Wallace as his running mate; his demand that party convention delegates go along with his decision changed the way vice presidents were chosen yet again, taking the power away from party leaders. Since 1960 almost every presidential candidate has named his running mate either at his party's national convention or a few days before. In the event that the president died or was incapacitated for any number of reasons, the Constitution did not make succession clear. An August 1961 *Chicago Tribune* headline, "KENNEDY PICKS JOHNSON TO BE HIS STAND-IN," suggests just how tenuous the arrangement was at the time. Kennedy and Johnson had an agreement that if Kennedy became incapacitated and could no longer serve, he would tell Johnson, who would then assume the presidency. If Kennedy was unable to communicate with Johnson, then Johnson could assume the powers of the presidency after consulting the Cabinet. The president could then decide at any time when he would be ready to take up the powers of the presidency again. It seemed like a haphazard way to make such an important decision.

Kennedy's assassination changed everything. On February 10, 1967, less than four years after Kennedy was killed, the Twenty-fifth Amendment to the Constitution was adopted. The amendment allows the vice president and Cabinet to remove the president from power if they believe he is incapable of holding the job, and it makes clear that the vice president will succeed the president: "Section 1. In case of the removal of the President from office or of his death or resignation, the Vice President shall become President." Under Section 4 of the Twenty-fifth Amendment, the vice president and a majority of either Cabinet officials or "such other body as Congress may by law provide" may declare in writing that the president "is unable to discharge the powers and duties of his office." The amendment places the power to replace the president squarely with the vice president—unless the vice president agrees that the president is unfit to serve, he cannot be removed.

Such a declaration would be submitted to the Senate president pro tempore, who presides over the Senate in the absence of the vice president, and the Speaker of the House. If the president objects, which he probably would, then a two-thirds vote by Congress would sanction his removal. If a vice president dies or is unable to serve, the amendment states that the president must name a successor to be confirmed by Congress (after Kennedy's death, Johnson did not have a vice president until he won the presidency himself in 1964).

But Cheney still does not think the wording is clear enough. In March 2001 he asked his loyal general counsel, David Addington, "to make absolutely certain that we're squared away because my main job here is to take over if something happens." Cheney said Addington reviewed all the procedures, statutes, and precedents, and the one very serious concern he had was that there was no procedure for removing the vice president in the Twenty-fifth Amendment. "Dave and I were concerned, especially given my health history." By that point Cheney had already had four heart attacks, his first when he was just thirty-seven years old running for Congress in Wyoming, and his fourth while he and George W. Bush waited for the Florida recount in November 2000. He thought of President Woodrow Wilson, who was incapacitated by a major stroke while he was in office and who left his second wife, Edith Bolling Galt Wilson, to run the country for well over a year. An added concern was that, in order for the Twenty-fifth Amendment to be carried out, the vice president would have to be part of the effort to get rid of the president. An incapacitated Cheney would mean that an incapacitated Bush could not be replaced. "I thought it was very important that we not get into the situation where I was in the office but I was incapacitated and unable to function," Cheney said, "so we came up with the idea of the signed resignation." He told Bush about his plan to write a resignation letter. Bush was a little taken aback by the idea but thought it made sense, given Cheney's health issues. On March 28, 2001, Cheney took out his stationery with the vice presidential seal and wrote: "Dave Addington—You are to present the attached document to President George W. Bush if the need ever arises.—Richard B. Cheney."

He told Addington that it would be Bush's decision alone if the letter should be delivered to the secretary of state. "I won't give specific instructions about when this letter should be triggered, but you need to understand something. This is not your decision to make. This is not [his wife] Lynne's decision to make. The only thing you are to do if I become incapacitated, is get this letter out and give it to the president," Cheney instructed Addington. "It's his decision, and his alone." Addington slipped the resignation letter inside two manila envelopes and hid it in a dresser drawer at his home—where he thought it would be safer than in the White House. In a twist of fate, his house was destroyed by a fire a few years later and, after getting his family safely out of the house, Addington retrieved the folder.

———

The vice presidency has expanded from understudy to true presidential partner. Presidential candidates now look for someone who can complement them in office, not someone who simply comes from a different part of the country and can balance the ticket geographically so that they will be elected. "The rapid evolution of technology, communications, and speed of travel has reduced many of the perceived ticket balancing needs of the old politics from the nineteenth and twentieth centuries," said Jimmy Carter, whose vice president, Walter Mondale, was picked in part because he was from the north and could balance Carter, a Southerner. "I believe personal compatibility, the balance of task assignments, and political and intellectual skills needed for governing in the complex world of today are much more important."

Contrary to the experience of fictional vice president Selina Meyer, who, in HBO's Emmy-winning show *Veep,* always got a resounding "No" when she asked her secretary if the president had called, the reality can be remarkably different.

"I worked for two vice presidents who spent most of their days, every day with the president," White House adviser Klain said. "They didn't have to sit by their phones and wait for a call." But no matter how close vice presidents are to the presidents they serve, they cannot escape the reality that they are guests in the White House.

II

Two Men, Two Hotel Suites

Shit, shit, shit.

—BOBBY KENNEDY, AFTER HIS BROTHER TOLD HIM HE WAS
PICKING LYNDON JOHNSON AS HIS RUNNING MATE.
NINTH FLOOR, BILTMORE HOTEL, LOS ANGELES, 1960

Now, where the hell's Bush?

—RONALD REAGAN, SEARCHING FOR A NEW RUNNING MATE
AFTER GERALD FORD TURNED HIM DOWN.
SIXTY-NINTH FLOOR, DETROIT RENAISSANCE CENTER HOTEL, 1980

When John Kennedy was a young senator recovering from a serious back operation, he wrote a letter to then–Senate Majority Leader Lyndon B. Johnson requesting a spot on the coveted finance, appropriations, or foreign relations committees. Johnson sent him a note: "It has been many years since I have enjoyed working with anyone as much as I have with you." But three days later Kennedy received another letter from Johnson, this one much less collegial: Kennedy would not be getting a seat on any of the committees he requested. Johnson's power over him at that moment was absolute, and he intended to use it.

Years after that exchange, much had changed. In 1960, Kennedy, forty-three, overwhelmingly won the Democratic Party's nomination and Johnson, fifty-one, was his main opponent. Johnson, who was a looming presence in the campaign even though he had not participated in the primaries, came in second in delegate voting at the convention. Before he won the Democratic nomination, Kennedy told his aides that he would name a "Midwestern liberal" as his running mate. The other two men who were challenging him for the Democratic nomination, Minnesota senator Hubert Humphrey and Missouri senator Stuart Symington, were at the top of Kennedy's list. But Johnson had not only come in second in the balloting, he had also won the South by a large margin, and no Democrat had ever been elected president without carrying the South. With only twenty-four hours to pick a running mate, Kennedy had to reconsider. If elected, Kennedy would be the first Catholic president at a time when many Americans feared that a Catholic president would take orders from the Pope and disregard the constitutional separation of church and state. In order for a Catholic to win, Kennedy had to carry the South, and he felt that he needed to bring Johnson on board—Johnson's home state of Texas alone had twenty-four electoral votes.

Both men were staying in Los Angeles's elegant Biltmore Hotel, where a delicate dance unfolded as the selection of a vice president consumed Kennedy and his younger brother, Bobby, who ran his campaign. Kennedy, Johnson, and their respective campaign teams were in matching large corner suites two floors apart in the art deco hotel, each with three main rooms and a series of standard hotel bedrooms stretching down a poorly lit corridor—Kennedy's on the ninth floor and Johnson's on the seventh. In both suites, the doors to the bedrooms were kept open and the atmosphere felt more like that of a standard company office than a luxury hotel.

Johnson and Kennedy were from two different worlds—Johnson had grown up Protestant and dirt poor in Texas, Kennedy was an Irish Catholic New Englander and the child of Joe Kennedy, the incredibly wealthy former U.S. ambassador to the United Kingdom. Johnson graduated from Southwest Texas State Teachers College (now Texas State University), Kennedy from

Harvard. Johnson was unrefined and unaware of the absurdity of some of his remarks (for instance when John Kennedy Jr. was born, Johnson sent Kennedy a telegram: "Name that boy 'Lyndon Johnson' and a heifer calf will be his"). Johnson was six feet, three inches and a powerful Senate leader who had a reputation for looming over colleagues, an act of physical intimidation so famous that it became known as the so-called Johnson treatment. Kennedy was much more sophisticated and reserved, and so were most of the people he surrounded himself with. As Kennedy's personal secretary Evelyn Lincoln would later write of the Johnson and Kennedy aides, "It was almost as if Romeo and Juliet had gotten married, and now Mr. Montague must somehow try to get along with Mr. Capulet."

But the union almost did not happen. Kennedy always thought Johnson would reject an offer to join his ticket, but the night before, when Kennedy won the nomination, Johnson sent him a telegram that suggested otherwise. It read, "LBJ now means Let's Back Jack." Johnson was not-so-subtly signaling his interest. He had already asked his aides to research how many vice presidents had become president (ten at that point) and to report back with the number of presidents who had died in office (then seven). At around 8:00 a.m. on Thursday, July 14, 1960, the morning after the Massachusetts senator won the Democratic nomination, Kennedy called Johnson's suite. Johnson's wife, Lady Bird, picked up the phone and shook her husband, who was sleeping in the twin bed next to hers. Kennedy told Johnson he wanted to talk with him in private, and they agreed he would come down to Johnson's suite at around 10:00. Johnson jumped out of bed and told a secretary to tidy up the suite's living room—he knew what this meant. Then he called his closest advisers, the future Texas governor John Connally, and his campaign manager James Rowe.

"Kennedy is coming down here in a few minutes," Johnson told Rowe, "and I think he's going to offer me the vice presidency. What should I do?"

"What do you want that for? You've got the power now," Rowe replied, referring to Johnson's position as Senate majority leader.

But Johnson was interested. "Power is where power goes," he said. He estimated his odds of ascension to the presidency if he became vice president to be approximately one in four. He is said to have told the politician and diplomat Clare Boothe Luce, "I'm a gambling man, darling, and this is the only chance I've got."

Before the call, the Kennedy brothers had huddled and, in hushed voices, John told Bobby that he wasn't going to choose labor union leader Walter Reuther for vice president, as Bobby had hoped. Several Southern governors had visited him and said he needed Johnson on the ticket to win. Bobby was furious—"Shit, shit, shit," he muttered as he left his brother's suite. Bobby hated Johnson, and that hatred clouded his thinking and could not be masked. Other Kennedy aides were just as apoplectic. "I was so furious I could hardly talk," Kennedy confidant Ken O'Donnell recalled. "I thought of the promises we had made to the labor leaders and the civil rights groups, the assurances we had given that Johnson would not be on the ticket . . . I felt that we had been double-crossed."

O'Donnell insisted on talking to Kennedy himself. Bobby brought him up to the ninth-floor suite and when Kennedy saw O'Donnell's devastated face he told him that the offer was only pro forma and that maybe Johnson wouldn't accept, and if he did there was an added benefit of keeping him off Capitol Hill. "I won't be able to live with Lyndon Johnson as leader," Kennedy reasoned. "Did it occur to you that if Lyndon Johnson becomes vice president, I'll have Mike Mansfield as the leader. . . . Somebody I can trust and depend on."

Johnson's loyalty to Kennedy would be tested from the very beginning during the humiliating behind-the-scenes ordeal at the convention. Johnson had said that he could not accept the offer without the blessing of his fellow Texan and mentor, Speaker of the House Sam Rayburn. But Rayburn was against Johnson being vice president, and he told Johnson and his wife that he "would not be happy" without him on the Hill. Rayburn knew that Johnson had wanted to be president ever since he was a young boy, and he knew he would not be content with being number two. Rayburn was all too aware of the terrible experience of FDR's first vice president, John Nance Garner—in fact, he considered Garner a mentor and had seen what happened after his friend, who was himself Speaker of the House, accepted Roosevelt's vice presidential offer in 1932: Garner became so bitter and vengeful toward Roosevelt, who he likened to a "dictator," that he became part of a "Stop Roosevelt" movement when FDR sought an unprecedented third term. But maybe this would be different. Now, Kennedy personally visited Rayburn in his hotel suite, promised to treat Johnson well, and persuaded Rayburn to support the idea, with caveats.

"I'm dead set against this but I've thought it over, and I'm going to tell you several things," Rayburn told Kennedy. "If you tell me that you have to have Lyndon on the ticket in order to win the election, and if you tell me that you'll go before the world and tell the world that Lyndon is your choice and that you insist on his being the nominee, and if you'll make every possible use of him in the National Security Council and every other way to keep him busy and keep him happy, then the objections that I have had I'm willing to withdraw."

But the Kennedy-Johnson alliance would always be difficult, in part because of Bobby Kennedy. Bobby more than disliked Johnson—he called him "mean, bitter, and vicious . . . an animal in many ways." It began when Bobby was a young aide to the powerful Wisconsin senator Joe McCarthy. He had heard Johnson delight in telling a story that made fun of his father, family patriarch Joe Kennedy. As Johnson told it, when he was a young Texas congressman, President Franklin Roosevelt summoned him to the White House and told him of his plans to fire Kennedy, who was then ambassador to the United Kingdom. Bobby idolized his father and was deeply offended that Johnson would make light of his father's dismissal. Johnson and Bobby first met in the Senate cafeteria in 1953, when, one morning during breakfast, Johnson walked by the twenty-seven-year-old, who was sitting with McCarthy and several other aides. Everyone stood to shake Johnson's hand and pay homage to the powerful senator. Everyone, that is, except Bobby. Johnson aide Horace Busby described the look on Bobby's face as "sort of a glower." After Johnson shook everyone's hand he looked down at Bobby, who was still seated, and hovered over him until Bobby could not avoid standing and shaking his hand, too. But even then he refused to make eye contact. "He didn't want to get up," Busby recalled, "but Johnson was kind of forcing him to." To make matters worse, during the campaign Johnson implied that the Kennedy patriarch had been friendly to the "appeasement" government of British prime minister Neville Chamberlain when Kennedy was U.S. ambassador to Britain before World War II broke out. Johnson even said that Joe Kennedy thought "Hitler was right."

Also during the campaign, Johnson's decision to bring up John Kennedy's poor health—dispatching campaign aides to relay Kennedy's afflictions to the press during the primary—only deepened Bobby's disdain. Kennedy needed daily doses of cortisone because of his Addison's disease, a painful illness that results from damaged adrenal glands. Johnson described the older Kennedy brother as a "little scrawny fellow with rickets" and "not a man's man." Then he attacked Kennedy for his long campaign and said, "Jack was out kissing babies while I was passing bills, including his bills." Johnson made it clear that he considered Kennedy an entitled elitist who was too young and too inexperienced for the job. "Have you heard the news?" he asked a congressman. "Jack's pediatricians have just given him a clean bill of health!" Bobby vowed never to forget any of it.

Bobby spent three hours and made several feverish trips between that morning and Kennedy's midafternoon announcement of Johnson as his running mate, using the back stairs between his suite on the ninth floor and Johnson's on the seventh in a persistent but vain attempt to get Johnson to reject his brother's offer. At around 1:30 p.m. Bobby went to the suite wanting to see Johnson but Johnson refused. Rayburn and Connally instead agreed to sit down with Bobby. They saw the younger Kennedy in a state of sheer panic. "We've got to persuade Lyndon not to take this vice-presidential thing," Bobby begged. "I don't know why my brother made the offer, but it's a terrible mistake. There's a revolt brewing on the floor. Labor is off the reservation, the liberals are in revolt. You've just got to persuade him not to accept this." Bobby wanted to know if Johnson would be satisfied being chair of the Democratic National Committee instead.

"Shit," Rayburn replied and went to the bedroom where Johnson and Lady Bird were mulling over the confusing events of that morning. Rayburn asked if Johnson would agree to meet with Bobby. Lady Bird told her husband not to do it, that Kennedy needed to personally withdraw the offer himself. Although Bobby was not able to convince Johnson to bow out, he did cement a bitterness between the two men that would last until the end of each of their lives.

Finally, at about 3:30 p.m., Kennedy called Johnson, who was still sitting on his hotel room bed, and read him the press release announcing Johnson as his running mate. "Well, if you really want me, I'll do it," a confused and exhausted Johnson replied. Incredibly, about a half-hour later Bobby made one more visit to Johnson, begging him to reconsider, but Johnson again refused.

Bobby recalled later, "He is one of the greatest looking sad people in the world—you know, he can turn that on. I thought he'd burst into tears. He just shook, and tears came into his eyes, and he said, swallowing his pride: 'I want to be Vice President, and, if the President will have me, I'll join him in making a fight for it.'" Up until a few minutes before Johnson announced that he had accepted Kennedy's offer, Bobby was still pressuring him to say no. Finally, *Washington Post* publisher Philip Graham, who was in Johnson's suite part of the day, called Kennedy and asked him what he wanted Johnson to do. "Oh," Kennedy said, sounding completely calm, "that's all right; Bobby's been out of touch and doesn't know what's been happening."

"Well, what do you want Lyndon to do?" Graham asked.

"I want him to make a statement right away," Kennedy replied.

Johnson and Lady Bird, his dutiful wife, stood on chairs in the hallway outside their suite to make the announcement as reporters jammed tightly together. Kennedy had made a statement at a packed press conference a few minutes earlier. "We need men of strength if we are to be strong and if we are to prevail and lead the world on the road to freedom. Lyndon Johnson has demonstrated on many occasions his brilliant qualifications for the leadership we require today," Kennedy told a stunned crowd, many of whom gasped in disbelief at the pick. When Johnson was back in the privacy of his suite he railed against Bobby, using his customary colorful language, which included calling him a "little shitass."

Johnson's alliance with Kennedy marked the first time in American history that two sitting senators—who were at odds just the day before—were nominated president and vice president. Johnson's mentor Sam Rayburn was right to warn him against the vice presidency, though—he absolutely hated it. At a dinner party in 1961, two months after taking office, Johnson told his friend the journalist Drew Pearson, "You know there was only one reason why I am playing second fiddle and why Lady Bird is playing second fiddle and why I am not running the Senate anymore. It's because I didn't want Nixon to win. If I hadn't taken second place on this ticket, Nixon would have won."

Johnson began exacting his revenge before Kennedy even took office. On November 17, 1960, nine days after they won the election, Johnson invited Kennedy to his ranch along the Pedernales River near Johnson City, Texas. It was an invitation Kennedy could not refuse. The president-elect arrived late at night and Johnson in-

sisted on a guided tour to inspect his cattle and pay his respects to
the Johnson family gravesite. A Kennedy aide described the noc-
turnal expedition as payback for what Bobby put Johnson through
at the Biltmore Hotel. Then Johnson woke Kennedy up early the
next morning to go deer hunting. Kennedy, who did not like to
hunt, looked directly into the eyes of a deer, according to his wife,
Jacqueline, fired his gun, and ran back to his car. Jacqueline told
the writer William Manchester that the visit forever haunted her
husband.

———————

Twenty years later, in another hotel, another sort of shuttle diplomacy took place as one aspiring Republican president tried desperately to broker an unprecedented deal with a former president
who was staying just one floor above him in the Detroit Renaissance Center Hotel during the Republican National Convention.
As with Johnson and Kennedy, the last-minute negotiations and
frenzied back-and-forth had to do with the vice presidency.

California political consultant Stuart Spencer had just signed on
with Ronald Reagan's 1980 campaign. On the plane traveling to
the convention in Detroit that July, according to Spencer, Reagan
spent the first half hour "dumping all over [George H. W.] Bush."
He was finding it difficult to forgive Bush for calling his tax policy
"voodoo economics" during the primaries. "Reagan was not the
kind of guy to get mad, but boy did voodoo economics really stick
in his craw," Spencer recalled.

"What do you think about a VP pick?" Reagan asked Spencer
after railing against Bush.

"Well," Spencer said, "I think you should pick George Bush."
Bush had finished second in the primaries, and polling data showed
that he could help Reagan beat Jimmy Carter.

"Haven't you been listening?" Reagan snapped.

There was another, more interesting pick Reagan was considering, someone who one poll suggested could move the ticket by

eleven points, much more than any other vice presidential prospect. This person, though, had been vice president and president, and it would take a very generous offer to get him to consider stooping so low.

When he was the sitting president Gerald Ford had resented Ronald Reagan, then a former governor of California, for taking the bold action of challenging him for the Republican nomination in 1976 and attacking him for being "too liberal" and too soft on the Soviet Union. Reagan's challenge, Ford thought, distracted him from the general election race against Carter. Ford called the Republican convention in Kansas City, Missouri, that year a "bloodbath." After Ford secured the nomination he refused to consider Reagan as his running mate, even when polls showed he was the smartest choice. Reagan, it turns out, would not have taken the job. He only agreed to meet with Ford after his defeat if he could be guaranteed that Ford would not offer him the vice presidency. Now, ironically, Reagan was begging Ford to join his ticket. No president who had been elected would ever accept the vice presidency, but maybe someone who had never been elected president, like Ford, might consider it. (Ford was Nixon's vice president and he became president after Nixon resigned over Watergate.)

In June 1980 Reagan had called his old rival and asked if he could pay him a visit at his home in Rancho Mirage, California. During that meeting Reagan offered Ford the vice presidency. Ford, Reagan reasoned, had the credentials he was lacking. One aide referred to the unlikely duo as a "dream ticket." But Ford declined Reagan's offer; he was thinking of running for president himself again. Dick Cheney, who was Ford's chief of staff in the White House, recalled a meeting with Ford and a handful of advisers to discuss Ford's own potential candidacy in 1980. They spent an entire day going over the pros and cons of whether Ford should run for president. The next morning the former president called them into his office with an answer: "Guys," Ford said,

"I've been thinking about it. Starting tomorrow you're going to be in here by noon and you're going to have a list with all the things we have to do. Frankly, I don't want to do it. Deep inside, I'd like to be president again but I'm just not interested in what I'd have to do."

But Reagan and his wife, Nancy, were not giving up. A month later in Detroit the Reagans visited Ford and his wife, Betty, in their hotel suite. At the convention Ford was thrilling Republicans with his blistering attacks on Carter. "You've all heard Carter's alibis," he told the crowd on the first night of the convention. "Inflation cannot be controlled. The world has changed. We can no longer protect our diplomats in foreign capitals, nor our working men on Detroit's assembly lines. We must lower our expectations. We must be realistic. We must prudently retreat. Baloney!" The crowd burst into applause; Ford was looking better and better. During their visit to the Fords' suite, the Reagans offered them an elegantly wrapped gift—an Indian peace pipe. "Ron," Ford said, "I thought the matter had been settled." Reagan pleaded with him, "Would you take another look at it?" Ford relented and said, "In deference to your request, I will." And so, on the third evening of the Republican convention, and on the very night of Reagan's nomination, a back-and-forth began between a man who had served as both vice president and president and a newly minted presidential nominee determined to win him over.

At the convention Ford said: "Elder statesmen are supposed to sit quietly by and smile wisely from the sidelines. I've never been much for sitting. I've never spent much time on the sidelines." Here was his chance to get back in the game. As he was mulling over the offer, he told journalists, he could imagine his partnership with Reagan being like the one in some European countries "where you have a head of state and a head of government." The question was, how much was each man willing to give up.

Spencer took his old friend Betty Ford aside. She was one person who had made up her mind about Reagan's offer. Betty was a famously outspoken first lady who was now finally enjoying time alone with her husband and family in southern California. "I don't know what's going on but he's looking for a divorce if he does this," she told Spencer. Meantime, Hamilton Jordan and Jerry Rafshoon, two of Democratic opponent Jimmy Carter's most trusted aides, were on separate vacations and giddy with excitement as they watched the Republican convention on television. "We were hoping that Ford would go through with it because rather than it being an advantage, it would show that they knew that Reagan couldn't handle the job," Rafshoon recalled.

Ford knew the perils of the vice presidency well and he did not want to be "a useless appendage," so he and his aides came up with a broad list of demands. One of Ford's unprecedented requests was to be made Reagan's chief of staff. "If the Vice President was the chief of staff, he would know everything," Ford reasoned, "and if something happened to the President, he could step into those shoes without any problem whatsoever." Reagan campaign aides called Cheney, then a congressman and delegate to the convention, and Bob Teeter, a Republican pollster, and together, along with former secretary of state Henry Kissinger, they made up Ford's makeshift team and went through his extensive list. Later, Reagan's team presented Ford's staff with a one-and-a-half-page document outlining exactly what Reagan was willing to offer, and it was extraordinary: making Ford chief of staff and giving him control over the National Security Council, the Council of Economic Advisers, and the budget office. "It was amazing the extent to which they were going to give up parts of the presidency," Cheney recalled. "Reagan went a long way toward meeting that list. We're all scratching our heads thinking, *This isn't going to fly, and if it does it's delegating too much of the presidency.* My belief always was that Ford really did not want it, he figured if he asked for all of those big requirements he wouldn't get it."

Their agreement began to unravel just as quickly as it came into being. Reagan thought Ford was going too far when he demanded that Henry Kissinger, who had been secretary of state in the Nixon and Ford administrations, be appointed to the same position in a Reagan White House. "Jerry, I know all of Kissinger's strong points and there's no question that he should play a role. I would use him a lot but not as secretary of state, I've been all over the country the last several years and Kissinger carries a lot of baggage. I couldn't accept that. My own people, in fact, wouldn't accept it," Reagan said. Kissinger was one of Ford's negotiators going back and forth between the hotel suites, which only added to the awkwardness of the situation.

The final straw came when Reagan, in the quiet of his hotel suite, saw an interview Ford did with CBS News's Walter Cronkite. When Cronkite suggested that the power sharing agreement between the two men would look something like a "co-presidency," Ford did not flinch. "I have to have responsible assurances," Ford replied. A co-presidency agreed to in private was one thing; making it public was quite another. Reagan, who was usually calm and affable, was visibly disturbed by what he had seen. "Did you hear what he said about a co-presidency?" he asked his aides. It was at that moment that Reagan came to a realization: *Wait a minute, this is really two presidents he is talking about.* At 9:15 p.m., Reagan called Ford and said they needed to wrap up a deal that night. At 10:00 Ford's aides asked for another day to consider the offer and were told no, it was now or never. At 11:30 Ford changed into a business suit, took the short walk downstairs to Reagan's suite, and told him, "This isn't gonna work." There was no way he could go back to the vice presidency. It was a short but emotional meeting. The two men hugged and Ford promised to campaign for Reagan in the fall. At 11:35 Ford left the room and Reagan suddenly found himself in desperate need of another option, and fast.

During the plane ride to Detroit days before, when Stuart Spencer had told Reagan he should pick Bush, a man he barely knew, he made this argument: "You're going to the convention that has a right-wing platform, and you need someone perceived as moderate on the ticket with you," he advised him. "Hmmm," Reagan mumbled, unconvinced, and the subject was dropped.

At that time he had his heart set on Ford. But now everything had changed.

Reagan said aloud, to no one in particular, "Now, where the hell's Bush?" By 11:38 Bush was on the line.

III

The Art of the Vet

Several people didn't want to be vetted because of girlfriends . . . People know when they can't take a frisk—they know that. If they've been sleeping with their secretary for twenty years, they know that people know that.

—JIM JOHNSON, DEMOCRATIC LAWYER WHO VETTED VICE
PRESIDENTIAL CANDIDATES FOR JOHN KERRY AND BARACK OBAMA

This is kind of like having a colonoscopy without anesthesia.

—CONNECTICUT SENATOR JOE LIEBERMAN ON BEING VETTED
TO BECOME AL GORE'S RUNNING MATE IN 2000

After eight years of being Reagan's VP, George H. W. Bush ran for president and won the Republican nomination in 1988. He then surprised everyone, including his close friend and adviser James Baker, when he picked little-known forty-one-year-old Indiana senator Dan Quayle as his running mate. Bush hoped Quayle would help attract votes from baby boomers and Christian conservatives who thought he was too moderate. In the weeks leading up to the announcement, his closest advisers presented him with lists of the three strongest contenders, the names constantly changing as seasoned Republican senators shuffled on and off, including Kansas senator Bob Dole and New Mexico senator Pete Domenici. Quayle's was the one permanent name on the list, and he was the only young, telegenic option. Plus, he lobbied for the job. "As I look back on it," Dole recalled, "Dan spent a lot of time in Bush's office visiting with him. I think they developed a closer relationship than we knew."

In a 1990 interview, Gerald Ford said that Bush's decision to pick Quayle was his biggest mistake. But that was in part because Bush did not let even his closest advisers in on his decision. He held it so tightly that he only whispered it to Reagan as he was about to board the plane that August to take him to New Orleans, where he would announce his running mate at the Republican National Convention. On the flight to Louisiana, Bush aide Craig Fuller and other top advisers were still throwing out names of who they thought Bush would pick, and no one guessed Quayle, including Baker, who was one of Bush's closest friends and a frequent tennis partner. Even though Quayle was the one constant on the lists of candidates, he was too much of an unknown, too much of a wild card, they thought. But Bush wanted to show that he was capable of bold decision-making, and the vice presidential pick is one of the few decisions that truly belongs to the nominee. When Bush called Quayle the Monday after the Democratic convention the previous month and asked if he would consider being his running mate, even members of Quayle's family were surprised. One of his children said they thought Bush was going to pick Dole, and another said, "You're not even a famous senator yet."

"The only decision I would take back in a nanosecond was announcing Quayle the way we did," Fuller reflected. "It did not give Quayle time to collect his thoughts." The announcement was indeed dramatic, with Bush and his wife, Barbara, arriving at Spanish Plaza, a riverside dock, on board a riverboat on the Mississippi. The plaza was teeming with people on the sweltering August day, and no one on board the boat could spot Quayle on the dock, in part because so many of Bush's team did not even have a good idea of what he looked like; Secret Service agents had trouble recognizing him. Quayle and his wife frantically pushed through the crowd. Finally, South Carolina senator Strom Thurmond spotted the Quayles and told Bush's aides. "Dan and Marilyn had trouble getting to the platform because they looked too young and no one realized why they needed to be up there," Barbara Bush later joked. Once at the podium, Quayle grabbed Bush's arm and shouted, "Let's go get 'em!" Quayle later acknowledged that it might not have been the best move. "I was picked, in part, for my youth, but this was a little too youthful—not vice presidential nor, given the nature of that office, potentially presidential." With that, Quayle began his rocky relationship with members of the press, who would question whether he had dodged the draft, smoked pot, and gotten lousy grades. In short, from the moment he was announced, questions arose about whether he had the credentials or the temperament to be first in line for the presidency.

During the campaign, Quayle was watched over by the veteran California political consultant Stuart Spencer, who worked for Reagan and admits Quayle was done no favors by the way he was selected. "If you pick someone like that you have to leak it early, you have to let the candidate go through the vetting process through the media," Spencer said. The decision to keep Quayle away from the press only further frayed his relationship with reporters. Moreover, Quayle and Spencer did not work well together. By the time he left New Orleans, Quayle admitted, "I was finding it difficult to trust myself." But Spencer said he had to micromanage Quayle because he was not prepared. "I was doing a press interview on the campaign plane sitting at a table and he and his wife are on one side and the reporter on the other side. The reporter asked him, 'What's the favorite book you've read?' He turns to his wife, Marilyn, and asks her, 'What's the favorite book I've read?' At that point I said, 'Holy shit.'"

––––––

Vetting and politics are supposed to be separate. "Vetting means wrongdoing, it doesn't mean lack of a political match," said Jim Johnson, who ran the vetting for John Kerry in 2004 and for a brief time for Barack Obama in 2008. Though most presidential candidates start thinking about a running mate well before the nomination, long and bitter primary battles can be distracting. Vetters often have less than a month to dig up everything they can possibly find on a potential running mate. Many of the questions they ask the finalists are cringeworthy. Rumors of infidelity or financial impropriety are often the tip of the iceberg, and sometimes the lawyers decide to call off a vet early if broader outreach will do nothing more than embarrass the person being vetted.

As more women make it onto the list of vice presidential candidates, the mostly male, white-collar Washington and New York lawyers who do most of the vetting are aware they need to rethink their approach. Instead of asking the candidate whether he's ever paid for an abortion, now they have to directly ask a woman if she's ever had one. Or, instead of asking a man if he's ever sexually harassed anyone, now the question is tweaked and the woman being vetted is asked whether *she* has ever been sexually harassed. One experienced vetting lawyer said that he prefers it when a female colleague sits in the room with him when he is asking such personal questions. Often the most shocking revelations come up in conversations rather than on written questionnaires, because no one wants a paper trail.

Lawyers also have to vet the running mate's spouse. When Walter Mondale ran against Ronald Reagan in 1984, he knew he had to make a historic pick if he was to beat the wildly popular sitting president. Tom Bradley, the first African American mayor of Los Angeles, was on Mondale's short list, along with Dianne Feinstein, then the mayor of San Francisco, and New York congresswoman Geraldine Ferraro. "We weren't doing oppo [opposition research] on Tom Bradley's wife," Mondale's campaign manager Joe Trippi explained—she was a private person who hosted teas and played the part of the supportive political wife, she was not in charge of a business. "Therefore why would you be doing oppo on Geraldine Ferraro and Dianne Feinstein's husbands?" But after Mondale picked Ferraro, the first woman to ever be the running mate of a major presidential candidate, revelations surfaced about her husband's business dealings that badly damaged the campaign. "From that day forward it was proven that you have to go much further than looking at the person themselves," Trippi said. Before then, vice presidential candidates were all men with wives who were mostly not working.

With less than a month to uncover every detail about a person's life, and without the help of the FBI, the job of vetting is more private eye than white-collar attorney. Most people, no matter how much they say they do not want to be vice president, eagerly answer vetters' questions. But the digging goes far beyond interviews with candidates. Jim Hamilton, a well-respected Democratic lawyer who vetted running mates for Kerry, Obama, and Hillary Clinton, said there are four standards for vetting: thoroughness, confidentiality, expedition, and respect. He recognizes that there is an inherent tension between them. Occasionally, he'll tell his lawyers not to interview too many people, especially when they travel to the candidate's hometown and track down their eighth-grade teacher. The more people who are interviewed, the more likely there'll be a leak.

————————

When presidents consider what kind of person they want to serve alongside them, they look for someone who can help them get elected and who can step up to the presidency. Usually, the most important consideration is whether the nominee is ready to take over. But presidential candidates also want something much simpler: they do not want to be embarrassed. In 1972, South Dakota senator George McGovern picked Missouri senator Thomas Eagleton as his running mate. At a time when mental health was even less understood than it is now, Eagleton was forced to leave the ticket after just eighteen days when it was revealed that he suffered from depression and had undergone electroshock therapy. Since

that embarrassing episode the vetting process has become deliberate and methodical. *Is there anything you would not want to see on the
front page of the* Washington Post? is a standard catch-all.

The list of questions is now well over a hundred. Vetters collect
everything that the candidate has ever written, they sit down with
spouses, they ask for details of former relationships, they log into
private social media accounts, and they insist on tax returns. They
even look at the potential nominee's children's, and in many cases
grandchildren's, Facebook pages and Instagram accounts. They
want to know everything, even if it's not disqualifying. In order
to make sure embarrassing revelations do not see the light of day,
anything that is written, including notes about what the candidate said during his or her interviews with the lawyers, is either
destroyed or given back to the person being vetted. "There are
no written reports of the vets," said Hamilton. "The reports are
made to the nominee orally. I don't want any written record." At
the beginning of the vetting process lawyers put together a "black
book" on each contender on the long list. The book is compiled
from publicly available material and is part of what is referred to as
a "blind vet," because the people being vetted do not know they
are under consideration.

Donald Rumsfeld was vetted for the vice presidency in 1974
(when Ford picked Rockefeller), 1976, and 1980. "There was a
guy walking around with me from the campaign at the convention
with a phone to be called if I happened to be the person picked,"
he recalled. He said it got easier every time he was on the list of
so-called usual suspects. When Ford had him vetted when he ran
in 1976, Rumsfeld was, at the time, Ford's secretary of defense. "I
laughed it off," he recalled. "The last thing in the world you're going to do is pick a former congressman from Illinois and a former
congressman from Michigan [Ford was a Michigan congressman
for twenty-four years]. It didn't make any sense. I think he probably had me on there for backup."

Richard Moe helped Warren Christopher do the vice presidential vetting for Bill Clinton in 1992. Moe gathered a team of lawyers and assigned each of them to one of the potential candidates. He said, "We would just take the guy apart financially and every which way." Former Johnson aide Harry McPherson oversaw Gore's vetting. Other candidates included Senator Bob Graham of Florida (whose daily diary, in which he cataloged the most mundane details of his life, was considered a disqualifier because it was deemed too eccentric), Nebraska senator Bob Kerrey, Pennsylvania senator Harris Wofford, and Congressman Lee Hamilton of Indiana. Gore was a senator from Tennessee, a relatively small Southern state like Clinton's Arkansas, who had run for president in 1988 and lost the Democratic nomination. Though they did not know each other well, Clinton, Gore said, was deferential and sent him a message through two mutual friends in 1991 that if Gore decided to run again, Clinton would bow out. It is a surprising concession for Clinton, who so desperately wanted to be president.

On June 30, 1992, under the cover of darkness, a red Bronco with tinted windows picked Gore up from his Capitol Hill apartment and brought him to the Washington Hilton for a meeting at 11:00 p.m. with Clinton. Gore's aides were worried that he would not be at his best so late at night—unlike Clinton, Gore is not a night owl. And Gore said he had "zero expectation" that he would be asked to join Clinton's ticket. They were fellow baby boomers and came from adjoining states, so picking Gore would be doubling down instead of balancing the ticket. "I really and truly felt that it was kind of a courtesy," Gore said of that night. "We spent a good bit of the time with me recommending who he ought to pick." Every other meeting with a finalist lasted no more than an hour, but this one went on until 1:40 a.m., almost three hours after it began. "I remember coming away after the three hours thinking to myself that, to my surprise, I thought he might actually want me to run with him." Despite their close age and geographic base, Gore balanced Clinton in other ways: he was considered more conservative (he had voted for the first Gulf War), and he was a Vietnam veteran (Clinton's patriotism was called into question during the campaign because of his efforts to avoid being drafted). And, if his reputation was correct, Gore was straitlaced and devoid of Clinton's messy personal life.

Even though Clinton advisers James Carville and Paul Begala were not sold on Gore, it was clear that Gore and Clinton got along and shared a similar worldview. "I told him at the beginning of our eight-year journey that he would never have any reason to doubt my loyalty or to not feel comfortable in putting one hundred percent trust in me," Gore recalled. But Clinton needed to be sure about Gore. The morning after their late-night meeting, Clinton set about personally investigating Gore and called his friends in Tennessee. When a friend of Gore's told Clinton, "Al Gore will not knife you in the back," the deal was sealed.

McPherson, the prominent Washington lawyer in charge of Gore vetting, also made off-the-record calls to Gore's old friends, reporters, and congressional colleagues looking for anything lurking in his past that the campaign should know about. He went to Gore's Capitol Hill apartment for an interview. It was a particularly hot day in Washington and the windows were open. The day before McPherson had called a reporter he knew in Tennessee and asked if there was anything at all that he should know about Gore. The reporter mentioned a rumor that Gore may have had a brief affair during his presidential campaign. But that was all it was—a rumor. Still, McPherson had to ask. One of Gore's selling points was that he had no dirty laundry. "Al," he said during their interview, "I don't want to ask you this question, and I'm embarrassed to ask you, but you'll understand more than anybody that with Clinton undergoing the scrutiny that he is, that he really would be in a bad way if he has a running mate with the same problems of the same kind."

Gore answered before McPherson could finish the sentence. But the windows were open and a diesel bus roared by the moment

Gore replied. McPherson could not quite make out what he said. He thought he said, "There won't be any problems," but he was not completely sure. The question was so awkward, especially for someone who seemed as prudish as Gore, that McPherson could not bring himself to repeat it. There is a self-righteousness about Gore that made those kinds of personal questions even more difficult to ask. During the campaign, Gore was incensed when a reporter asked if he had ever smoked pot. He was riding in a car with the reporter, his wife, Tipper, and an aide at the time and he sat there speechless until he finally asked the driver to stop the car. He told the reporter to step outside with him and asked him if he really wanted to ask that particular question. He had been through this before. When he ran for president in 1988 Gore admitted to smoking pot while at Harvard and in the army and he wanted the matter to be closed for good.

––––––––

In 2000 George W. Bush, then governor of Texas, wanted his father's former secretary of defense, Dick Cheney, to run his presidential campaign. Cheney, who was then the CEO of Halliburton, the Texas-based oil and gas extraction firm, said no. Bush then sent his chief of staff from the governor's office to see if Cheney would agree to be considered as his running mate, and again Cheney declined. Finally, Bush called Cheney and asked him, if he would not agree to be his vice president, would he at least agree to run the search to help him find one. Cheney finally relented and agreed to do it. Even Lynne Cheney never thought that her husband would become vice president—she told a close friend how relieved she was when Cheney told her that he would be on the committee to select Bush's running mate, thinking that it would ensure that he himself was out of the running.

Weeks before the Republican National Convention at the end of July, Cheney went to visit Bush at his ranch outside of Crawford, Texas. The two men sat at Bush's dining room table and flipped through binders crammed with the most intimate details of the lives of the actual candidates up for the job. Cheney had put together an especially rigorous questionnaire that included nearly two hundred questions under seventy-nine headings. One of Cheney's specific questions was whether there was anything in a candidate's past that would make him "vulnerable to blackmail or coercion." Bush asked Cheney to set the binders aside. "The man I really want to be the vice president is here at the table," Bush told him. Cheney said nothing. The two men went out to the back porch and in the searing summer heat Cheney recited the long list of reasons why he was not a wise choice: his bad heart; his two drunk-driving convictions; his flunking out of Yale University; and his daughter Mary, who is gay, and could be a lightning rod in the Republican Party, which still defines marriage as a legal union between one man and one woman. Gerald Ford had told Cheney that being vice president was "the worst eight months of my life"—those words were surely ringing in his ears at that moment. Cheney said he would have to think long and hard about Bush's offer. "I always saw the vice president as expendable in a sense," he said years later.

And there was a housekeeping issue to attend to: Cheney would have to change his voter registration because both he and Bush were registered to vote in Texas, and the president and vice president cannot be registered to vote in the same state. Eventually, of course, he agreed to take the job, but only if he was given the power to overhaul it. Bush and Cheney made clandestine trips in the weeks leading up to the convention to interview other candidates. Cheney sat in the room quietly, knowing all along that the decision had already been made.

———

When Jim Johnson was running the vetting for Democratic presidential nominee John Kerry in 2004, he was particularly worried about two candidates who had inadequate medical records. One had been examined by a staff member who happened to be a doctor. That would not do. It was conveyed to the potential nominee, whom Johnson would not name, that he would be picked up the next morning and taken to a local hospital and examined by doctors who were not on his payroll. The health of the vice president is a critical part of the very secretive vetting process. When Indiana senator Evan Bayh was being vetted as Obama's running mate in 2008, he had to pay several thousand dollars for medical tests to find out why he was having such bad stomachaches. It turned out he had adult-onset gluten sensitivity. "They wanted to make sure I wasn't dying," he said.

In 2004, Democrat John Kerry's first choice for his running mate was Republican senator John McCain of Arizona. The two knew each other from Vietnam and from their years in the Senate. "It was clear that it wouldn't fly. McCain didn't want to do it, it was clear that it would have been too controversial in other quarters," according to a person familiar with Kerry's decision-making. "But if you said where was his heart when this was moving along, I would say McCain." In the end, the vetters gave John Edwards, then a charismatic young Democratic senator from North Carolina who had sought the nomination that year, a long look. They read every legal submission that Edwards made in his career as a lawyer, and they looked at every speech he had ever delivered. They investigated reports that he had had affairs but, in the end, they found nothing conclusive. An affair with former campaign aide Rielle Hunter would later derail his own presidential run in 2008.

—————

In the early summer of 2008 Barack Obama, then the Democrat's presidential nominee, held a first round of meetings with his advisers on Capitol Hill to focus on running-mate options. Jim Johnson initially oversaw the vetting process along with Eric Holder, who, at the time was Obama's top legal adviser, and Caroline Kennedy, the daughter of President John F. Kennedy. More than seventy top Washington- and New York–based lawyers volunteered to be on the team. "When you say to somebody, 'Help me choose the top candidate's running mate,' they make time for that," Johnson quipped. "A lot of lawyers love doing this because it's being a part of history," said Hamilton. "I have no trouble finding people."

The vetting team initially focused on twenty-three people. Obama asked for the so-called black book on each of the candidates. Vetters flew to Obama's Chicago campaign headquarters to present him with the books. But there were twenty-two, not twenty-three.

"I have twenty-two books for you," a person who was involved in the process, but who spoke on the condition of anonymity, told Obama.

"But we agreed to twenty-three," Obama replied.

"Yes," the person said, "and I decided not to bring the twenty-third."

"I wanted twenty-three," Obama, who is famously unemotional, said, sounding a bit annoyed.

"It would be very bad politics for you to have a book that had been prepared on one of the candidates," the person told him, "because it's got so much dirt in it." The vetting lawyers were so concerned about this one person that they were worried about even putting a binder together on him, fearing it could be traced back to the Obama campaign. The information was too scandalous. Obama, of course, wanted to know who it was. The room was packed with campaign staff, so the person told Obama they should speak alone in the hallway.

"So, who didn't you do a book on?" Obama asked.

"Mike Bloomberg."

Obama nodded his head, but this time he kept his poker face and returned to the room. Once the lawyers had started considering Bloomberg, then mayor of New York City who had left the Republican Party in 2007 to become an independent, they began investigating his financial services and media company, Bloomberg L.P. They were surprised by what they found. One *New York Times* article published on May 2, 2008, noted that more than fifty women were accusing Bloomberg L.P. of discriminating against pregnant employees. One former sales executive claimed that after she became pregnant, Bloomberg encouraged her to have an abortion because so many female employees were taking maternity leave. He adamantly denied the allegations and settled that case out of court for an undisclosed amount. There was no disputing it, though: Bloomberg was absolutely off the table.

Joe Biden was decided on very early—Obama, who had served less than a full term as a senator from Illinois, wanted him as his running mate even before he won the nomination. He had the experience that Obama was lacking, and he came from a working-class Catholic family and could appeal to skeptical Democrats in the Rust Belt, particularly in Pennsylvania and Ohio. Biden had been a senator for thirty-six years and chair or ranking member of the Senate Foreign Relations Committee and Judiciary Committee for more than a dozen years apiece. Rahm Emanuel, who later became Obama's chief of staff, asked the vetting lawyers, "Why are you making this so complicated? The obvious choice is Biden, so why not just do that and forget about the rest of it?"

But all that time in the public arena also meant that Biden carried baggage. He ran for president in 1988 and was forced to drop out of the race after delivering passages from a speech by British Labour Party leader Neil Kinnock without attribution. Kinnock lost the election, but his lyrical speeches and populist message appealed to Biden. In one commercial Kinnock asked rhetorically why he was "the first Kinnock in a thousand generations to be able to get to university?" And pointed to his wife, who was sitting in the audience: "Why is Glenys the first woman in her family in a thousand generations to be able to get to university? Was it because all our predecessors were thick?" At the Iowa State Fair, an important stop for any presidential hopeful, Biden said nearly the exact same thing, just swapping out names: "I started thinking as I was coming over here, why is it that Joe Biden is the first in his family ever to go to a university?" he asked. Then, pointing to his wife, Jill, he continued: "Why is it that my wife who is sitting out there in the audience is the first in her family to ever go to college? Is it because our fathers and mothers were not bright? Is it because I'm the first Biden in a thousand generations to get a college and a graduate degree that I was smarter than the rest?"

Biden explained that one of the reasons for the slip was his distraction with the Senate hearings for Ronald Reagan's Supreme Court nominee Robert Bork, but he admitted that part of it was also his arrogant belief that he "could talk my way through the most important campaign event of the summer." He often speaks in the third person and exaggerates his biography, including in 2007 when he said he was "shot at" in Iraq, which he later corrected to say, "I was near where a shot landed." But the good far outweighed the bad. Biden has a Bill Clinton–esque ability to connect with people, drawing energy from a crowd, resting his hand on people's shoulders, hugging them, and generally being less professorial than the man who picked him as his number two.

Biden had dropped out of the presidential race in January 2008, after coming in a distant fifth in the Iowa caucuses. At Bush's last State of the Union address, Obama asked Biden for his support. "If you win, I'll do anything you ask me to do," Biden told him. Obama replied, "Be careful, because I may ask you a lot." During the remaining primaries, Obama sought Biden's advice about twice a week and was particularly interested in Biden's perspective on foreign policy matters. "The only question I have is not whether I want you in this administration," Obama told him. "It's which job you'd like best." At one point, Obama asked him if he would be more interested in being secretary of state than vice president, an offer Biden said he seriously considered. When it came to the vice presidency, Biden was not immediately convinced. "It wasn't self-evident to me that being vice president would be a better job."

Biden got a call from Obama asking him if he would agree to be vetted when he was on the Amtrak train, traveling from Washington to his home in Wilmington, Delaware. It was a commute he made almost every weekday while he was in the Senate. He was not enthusiastic. "I never wanted to be vice president, I thought I could help him more as chairman of the foreign relations committee," Biden said. "I thought I could be his fixer in the Senate between Democrats and Republicans." Biden told Obama he would get back to him and he had a family meeting that night to discuss the offer. Everyone in his family told him he should say yes, especially his wife, Jill, who thought he would be home more as vice president than if he agreed to be secretary of state, a job that requires constant travel.

Before Obama approached him to join the ticket, Biden's ninety-year-old mother, Catherine Eugenia "Jean" Finnegan Biden, had asked him what kind of man he thought Obama was. Biden told her he thought he was a good man. His mother was at the family meeting that night as they were all gathered on the back porch of Joe and Jill's Delaware house. She had not said a word, which, Biden said, was unusual for his chatty mother, who he absolutely worshipped.

"Honey, you haven't said anything. What do you think?" Biden asked her.

"Joey, remember when I asked you what kind of man he was? And you said he was a good man."

"Yes," he said.

"And remember when you were fourteen that realtor sold a house to a black couple and all those people came out and protested and I told you not to go down there and stand with the people who bought the house, and you went down and the police arrested you and brought you home because they were worried about your safety?"

"Yes," he repeated.

"And remember you had that good job at Windybush swimming pool and you were a lifeguard but you wanted to work in the inner city, you wanted to be the only white employee on the east side of Wilmington, Delaware?"

"Yeah, Mom, I remember that," he said.

"Remember after the city was burned down and you had that job with that white-shoe law firm, the oldest law firm in Delaware and after six months you quit and became a public defender?"

"Yes, Mom," he said.

"Let me get this straight, honey. The first black man in history who has the chance to be president says he needs you to win and you told him no."

Son of a bitch, he thought to himself, *she's right*. Biden said it was at that moment when he knew he had to say yes and accept Obama's offer. When he agreed to be vetted, he said, "I knew that once I did that that I'd crossed the Rubicon, that if he wanted me I would do it."

Obama chief campaign strategists David Axelrod and David Plouffe went to see each of the three finalists: Biden, Indiana senator Evan Bayh, and Virginia senator Tim Kaine. They flew into Wilmington in a cloak-and-dagger mission trying to avoid the press so that Obama's pick would be a surprise, and arrived at Biden's sister's house. Biden's eldest son, Beau, and Jill Biden then picked them up and brought them to Biden's house. Biden walked in wearing a baseball cap and his signature aviator sunglasses.

"The thing I remember very clearly was him kissing Beau and saying, 'I love you, I may be by later to see the kids,'" Axelrod recalled. Once they were seated, Axelrod delivered some blunt criticism. "Obviously, Senator Obama is really interested in you. The one big concern is that we have to work in concert, and frankly the whole loquaciousness thing can be an issue." Biden nodded and proceeded to launch into "a two-hour oration," Axelrod said with a laugh.

"I ran for president because I thought I would be the best president," Biden said. "And I still think that. But Barack won and I think it's important that he succeed and I want to help him succeed."

Axelrod was impressed. "I thought that was disarmingly honest.

I took that well because I thought he was leveling with us. Every ten minutes or so he'd come up for air and he'd say, 'Am I making sense here?' And, in fact, he was brilliant in a lot of his observations. It was a tour de force, honestly."

Biden says he needed at least two or three hours alone with Obama to hash out details of how he saw the job. Biden talked to Jimmy Carter's vice president, Walter Mondale, and Mondale sent him his formal agreement with Carter and encouraged him to act as a general adviser, and not to request specific assignments. Biden also asked Gore for advice. On August 6, the Obama campaign snuck Biden into Obama's Minneapolis hotel suite, where they stayed up late discussing how each of them saw the role. Obama told him the vetting had gone well, particularly because Biden did not have complicated finances. "All these years," Obama teased him, "and you still have no money."

In his lifetime Biden had seen too many vice presidents fail, and he felt like he was at the peak of his career and he did not want to waste time. His best friend and adviser, Ted Kaufman, told him, "If you become vice president I'm not going to be the one who goes into your office in the morning and tells you, 'You know what you wanted to do today, you're not going to be able to do it because some junior staffer from the White House said they want you to fly to Indiana to do something else.' I don't want to sit there and watch you get upset." Biden knew the pitfalls and told Obama what he really wanted, which was "to be the last guy in the room." Biden says, "It wasn't a throwaway line. I meant it literally, not figuratively."

Biden and Obama agreed to five ground rules in a private written document: "JRB and BO have weekly unstaffed meeting; JRB can sit in on any BO meeting; JRB must have contemporaneous receipt of all paper—All printed words that go to BO go to JRB; JRB staff must be included in any meeting with their parallel BO staff; JRB will not have a portfolio, because he will be involved in everything."

One other thing, Biden told Obama: "I'm not changing my brand. I am what I am."

———

Eight years before Donald Trump put A. B. Culvahouse in charge of his VP search, Culvahouse was running McCain's vice presidential vetting team. And he was instrumental in McCain's unconventional and controversial selection of Alaska governor Sarah Palin. McCain, aware that he faced an uphill battle against Barack Obama, the first African American to win the nomination of a major political party, felt that he needed a game changer. Hillary Clinton never made it onto Obama's final list of vice presidential possibilities, according to a person familiar with his decision-making, and that made nominating Palin, who would be the first Republican woman running mate, even more attractive.

Culvahouse was given a long list of twenty-six possibilities by the campaign, more than twice as many people as would be on Trump's initial list. The people on the short list—McCain's close friend Senator Joe Lieberman of Connecticut, Minnesota governor Tim Pawlenty, former Massachusetts governor Mitt Romney, and Florida governor Charlie Crist—had personal interviews with Culvahouse and were given nearly a hundred written questions to answer. Bloomberg, who was originally on Obama's long list of potential running mates before he was deemed radioactive, was also on the long list of people who were vetted using public material by McCain's team. McCain's lawyers, unlike Obama's, prepared a report for the candidate on Bloomberg but he was not interviewed.

McCain saw the decision as an opportunity to show that he was a maverick, a politician who was unafraid to take a risk and pick someone outside the mainstream Republican Party. Before he settled on Palin, Lieberman was McCain's top choice. Lieberman had the added advantage of having already been vetted as a running mate when he was picked by Gore eight years earlier.

Gore's team had reviewed more than eight hundred legal opinions Lieberman rendered when he was attorney general of Connecticut in the 1980s. But he would be a risky choice—Lieberman was a Democrat-turned-independent who supported abortion rights. South Carolina senator Lindsey Graham, who is very close to both McCain and Lieberman, leaked his name to the press. Culvahouse said that hurt Lieberman's chances because it gave conservatives the opportunity to mobilize against him: "If they had announced Lieberman when they announced Palin I'm not sure that the conservatives could have organized that quickly against Lieberman."

McCain was left scrambling for a running mate. Sarah Palin, who was not nationally known and had been governor of Alaska for less than three years, was not even on McCain's long list of potential candidates. She came to his attention because the other choices were too predictable and not the game changers he and his campaign strategists thought he needed in order to win. In mid-August 2008, less than a month before the convention, McCain met in Aspen with several people on his short list. A pollster told him, "Senator, a middle-aged white guy doesn't do it for you. If you're going to beat Obama you need to shake it up, you need a minority or a woman." The Republican Party's bench in those demographics was very thin. His top aides, including Steve Schmidt and Rick Davis, urged him to pick Palin, because, like McCain, she was a "maverick."

McCain's vetting lawyers had just seventy-two hours to vet Palin, who was forty-four years old and just six years removed from being mayor of Wasilla, Alaska. It was one of the biggest gambles in modern politics. Culvahouse asked Palin the questions he had asked McCain's other candidates, including one that made sense before President Obama ordered the raid that killed Osama Bin Laden, the al-Qaeda leader behind the 9/11 terror attacks: "You're the acting president, the president just had surgery and the director of intelligence comes in and says they have a confirmed sighting of Bin Laden in the northwest territories of Afghanistan. He tells you that we have a plane overhead ready to take the shot, but there will be multiple civilian casualties. Do you take the shot?"

Palin replied, "Yes, I would take the shot because I'm the President of the United States, this is our archenemy who took the lives of three-thousand-plus Americans. And then I would get down on my knees and ask for forgiveness for the innocent souls whose lives I would be taking."

"Now that's a brilliant answer," Culvahouse said, his voice filled with admiration nearly a decade after that conversation.

Palin did not reveal the pregnancy of her seventeen-year-old daughter, Bristol, on the written questionnaire, but she told Culvahouse during a phone conversation, "You've got to know that my daughter is pregnant and I'm not sure she's going to get married to the father." It was thought to be a deal breaker among some in McCain's orbit, but John and his wife, Cindy, were less risk averse. One matter Culvahouse was particularly concerned about was so-called Troopergate, which centered around Governor Palin's decision to fire Alaska's public safety commissioner Walt Monegan because he would not fire Mike Wooten, a state trooper who was going through a messy divorce and custody dispute with Palin's sister. Palin, Culvahouse said, had so much charisma that it overshadowed their concerns. "She had immense presence, she filled the room," he recalled.

McCain had only met Palin once, at the National Governors Association six months earlier, and they had only spoken for fifteen minutes. They spoke once by phone before Palin and her husband were flown down to the McCains' home near Sedona, Arizona, where McCain formally made the offer. When he announced Palin as his choice the day after that meeting, McCain had spent less than three hours with her. Palin's personal issues and her lack of exposure to national media attention led to a caricature of her as an unsophisticated and inexperienced vice presidential candidate. Campaign aides understand that vice presidents cannot help their candidate get elected, but they can certainly hurt. Palin proved that to be true. Palin cost McCain 2.1 million votes, or 1.6 percentage points, according to a 2010 study by researchers at Stanford University. Culvahouse is not apologetic about the decision, though, and says that Palin was "plenty bright." He thought she could get up to speed. But during a moment of reflection he admitted, "In retrospect, were we all dazzled by the force of her personality? That was part of her political appeal, she was a real force."

————————

Trump had several women on his long list, including former secretary of state Condoleezza Rice; South Carolina governor Nikki Haley, who is now ambassador to the United Nations; and Iowa senator Joni Ernst. Rice and Ernst took themselves off the list—Ernst bowed out because she thought she would not be picked and Rice, a perennial favorite, had taken herself off lists twice before, in 2008 and 2012. The biggest sticking point for Rice was Trump's position on the Iraq War and his criticism of the Bush administration in which she had served—during the campaign Trump said the invasion of Iraq was based on a "lie." No women made it onto Trump's short list, which was an eclectic collection of men who hold very different worldviews: Alabama senator Jeff Sessions, later named Trump's attorney general; New Jersey governor Chris Christie; Tennessee senator Bob Corker; Ohio governor John Kasich; former Speaker of the House and Republican firebrand Newt Gingrich; and of course Indiana governor Mike Pence. According to someone familiar with the process, Corker took himself out of the running at the very end because he was so conflicted about working for Trump, and he knew that if he agreed to be his vice president, Trump would expect absolute loyalty. In the final days, Gingrich and Pence, two very different men with very different backgrounds, were Trump's finalists.

Kasich, who challenged Trump in the Republican primaries and whose poll numbers Trump regularly mocked, was surprisingly on the list, no matter how many times he insisted he was not interested in the job. Kasich's campaign manager, John Weaver, provides a hint at how much power a vice president in the Trump administration might actually wield, or at least how much Trump's tight-knit family *hoped* he would wield. Weaver exchanged several text messages with Donald Trump Jr. in May 2016, shortly after Kasich dropped out of the race. "He asked to meet me at Trump Tower, which I told him I couldn't do. As Kasich's strategist, I can't just waltz into Trump Tower, so he called me while I was at home in Austin and made the offer," Weaver recounted. The offer was incredible, presenting a co-presidency not unlike what Reagan offered Ford during their negotiations in Detroit. "Kasich's the perfect choice," Trump Jr. told Weaver. "My father would really like him to do it. He would be the most powerful vice president in history." Weaver, skeptical that his boss and Trump, who had developed a visceral dislike for each other, could ever make it work, told Trump's son, "I've been around the block a few times, I've heard that before." (Trump's son-in-law, Jared Kushner, uses similar flattery in job interviews. According to one person who Kushner interviewed for a White House job, when asked exactly what his job description would be, Kushner simply told him: "You'll be in charge of everything.")

Trump Jr. further unnerved Weaver when he said, "You don't seem to understand, he would have control over domestic and foreign policy."

A puzzled Weaver replied, "Well, that's interesting, and what would the president be doing?"

"My father would be making America great again," Trump's eldest son assured him.

Weaver was shocked at how much Trump was willing to delegate to his number two. Trump Jr. was generally affable during the call, Weaver said, but he was "surprised that we didn't seem to grasp what an earth-shattering offer this was and he didn't understand why we didn't jump at it." Kasich did not even attend the Republican National Convention, which was held in his home state, because of his deep disdain for the Republican nominee, but up until the end Trump clung to the hope that he would reconsider, according to a person familiar with Trump's deliberations. It was not going to happen: Kasich did not even vote for Trump; instead he wrote in McCain's name. Weaver says he himself threw up for three hours straight on election night.

Trump Jr. was courting someone else at the same time. He is close with Fox News host Sean Hannity and aggressively pushed his father to pick Gingrich, a Fox News contributor and Hannity friend. (Hannity even provided a private plane for Gingrich to fly to Indianapolis when Trump was meeting with Pence.) But Gingrich had some major drawbacks. One top Republican state party chairman said that when Trump asked him what he thought—as he did nearly every Republican he encountered—he told Trump: "Newt is a brilliant person, a brilliant mind, but I do not believe that you need another lightning rod at the top of the ticket." Like Trump, Gingrich had been married three times, and he still owed his failed 2012 presidential campaign more than $4 million. A debt that large raises questions about what kind of leverage someone who helps him pay it off might have over him. Also, Gingrich's wife, Callista, insisted on continuing to run Gingrich Productions, a Virginia-based multimedia production company that produces documentaries such as *Divine Mercy: The Canonization of John Paul II* and *Rediscovering God in America.* "*She* wouldn't let that one go," said a person with direct knowledge of Trump's vetting process. A second lady raising money for a private enterprise is unheard of, and, as with their campaign debt, it would also beg questions about the leverage benefactors might have over the Gingriches while in office. (Callista has since been confirmed as Trump's ambassador to the Vatican.) Ethical questions remain surrounding Trump's own ties to his family real estate business while he is in office. An issue with the Gingriches would only compound the scandal.

New Jersey governor Chris Christie, who ended his campaign for the presidency in February 2016 and was one of the first major

politicians to back Trump during the primaries, was fully vetted.
But the so-called Bridgegate scandal, involving the deliberate clos-
ing of lanes on the George Washington Bridge linking Fort Lee,
New Jersey, and Manhattan in 2013, derailed Christie's chances.
The lane closures led to severe traffic jams and were an effort by
Christie allies to retaliate against the Democratic mayor of Fort
Lee because he did not endorse Christie's reelection bid. "The
timing of the Bridgegate trial was just impossible," said Trump's
vetting lawyer Culvahouse, who noted that the trial of two former
Christie associates was scheduled to begin in September, just two
months before the election. The Trumps were consumed with the
story. "The Trump family, Trump himself, they read the *New York
Times*, the New York media," Culvahouse said. "And it had been
all Bridgegate, all the time for a couple of years." It did not help
Christie that Trump's son-in-law, Jared Kushner, resented Christie, who, as U.S. attorney for New Jersey back in 2005, had played
a role in sending his father, billionaire real estate developer Charles
Kushner, to prison for tax evasion and retaliating against a federal
witness. "He was knifed in the back by Trump insiders," said a
Christie aide. "But he and Trump have a very good relationship
and Trump asks him for his advice often." The aide said his boss
has turned down a half-dozen other offers to serve in Trump's
administration.

––––––––

Toward the end of his three-week-long hunt for a vice president, a
flat tire grounded Donald Trump's plane in Indianapolis. Rumors
circulated that Trump's then–campaign chairman Paul Manafort
made up the story to keep Trump in town so that Indiana governor Mike Pence could make his case for the vice presidential
nomination. Pence certainly took advantage of the opportunity.
Trump and his son Eric were in town for a rally the night before, when Pence effectively auditioned for the role of running
mate, and while stranded they had dinner with Pence and his wife,

Karen, at an Indianapolis restaurant. At dawn the next morning, Trump's other two eldest children, Don Jr. and Ivanka, who, with Eric, formed his loyal kitchen cabinet during the campaign, flew in. They were concerned that their father might make the wrong decision and pick Gingrich, or even Christie, so they came in time for breakfast hosted by the Pences at the governor's mansion. During the meal Pence summoned some of what he had learned as a conservative radio host and railed against Hillary Clinton and the scandals that plagued the Clintons in the White House. It was a soothing sign for Trump's family members, who wanted to make sure that Pence would gel with Trump's aggressively anti-Hillary team. They were also working against some of Trump's top aides, including Steve Bannon, the antiestablishment agitator who would become Trump's chief White House strategist during his first several months in office and who was less than enthusiastic about Pence. According to leaked emails, Bannon called a Pence pick "an unfortunate necessity."

There was no great fanfare at the governor's residence. According to Pence's gubernatorial chief of staff, Jim Atterholt, "People assume you'll have a chef with a professional staff. Well, there's no chef at Indiana's governor's residence, there's no professional staff." Atterholt said Mike and Karen picked fresh flowers for the table, heated up breakfast, set the table themselves, and served the Trumps. "It was all very natural, it's what they do all the time," said Atterholt, to whom Pence talked privately about the breakfast. "I think the Trumps were fascinated by that." The fact that the Pences served breakfast and picked up the dishes afterward was foreign to members of a family who are used to an army of household help. The Trumps were mesmerized by the normalcy. "I can't convey to you how powerful that was," Atterholt said. "The humility they showed, they were just fascinated by them." Trump has said privately how incredible it is that Mike and Karen always seem to be holding hands (he and Melania rarely do in public). The Pences, in turn, were fascinated by Trump, a billionaire businessman with a glamorous wife who used to be a model. It was an odd chemistry that just happened to work.

Jeff Cardwell, chair of the Indiana Republican Party, recalled Trump grilling him about Pence at a fund-raiser days before he was picked. Trump said, "I understand you've known Mike Pence a long time." By then Trump had narrowed it down to Pence and Gingrich. He asked Cardwell why he should pick Pence—he wanted to know about his track record in Indiana. Cardwell listed the years Pence served in Congress and talked about his solid relationships with Republican governors. Finally, Cardwell told Trump, "He has great relationships with governors in the Midwest, like Rick Snyder in Michigan and Scott Walker in Wisconsin. No Republican is going to win the election without winning the Midwest." His words proved prophetic.

Pence was not the most obvious choice. During the campaign he called Trump's proposal to ban all Muslims from entering the United States "offensive and unconstitutional." And in the days leading up to the Indiana Republican primary there was great internal debate in the governor's office as to whether he should endorse anyone. His staff, almost to a person, was unanimous that he should stay out of it. "We could tell that Trump was very popular in Indiana, but the governor for whatever reason, pretty much on his own on this as best as we could tell, felt like he should endorse Senator Cruz because he saw him as a principled conservative," said Atterholt. "He really wasn't anti-Trump, he just felt that in order to stay true to his conservative principles he had to endorse Cruz." Trump even had a meeting with Pence in April, facilitated by Chris Christie, in an effort to keep him neutral, but Pence would not budge.

Pence went on a local radio show before the Indiana primary and endorsed Cruz, but he devoted much of the time talking up Trump and praising him for giving voice to working men and women in the country. Later Trump told Pence privately, "I'm used to fistfights in politics, I've never seen somebody endorse somebody else and say something kind about the other person at the same time." Once again, Trump was fascinated by Pence—he'd assumed that Pence was just going to trash him and he was totally prepared for it—and Pence's generosity left him dumbfounded.

Pence was in the right place at the right time. In fact, if Trump had not picked him as his running mate, polls show he might not have won reelection as governor. There was some concern about his years as a radio host, though. Trump's lawyers could not find most of the tapes from his radio days and privately wondered whether they had simply been recorded over or were purposely destroyed. Ultimately, Trump's lawyers decided that nothing Pence said in his past was cause for real concern. Pence's wife had had a short first marriage, and that was a brief consideration (Jill Biden had been married once before, too). There was some pressure on Trump to make up his mind because Pence had to decide whether he was going to run for reelection as governor and the deadline to remove his name from the ballot was looming. "We certainly were not going to file to take his name off the ballot for reelection until we were certain he was the pick," Atterholt said. "It came down to the last couple of days. I think that the filing deadline accelerated the overall selection process for Trump."

On Friday, June 10, the day before the Indiana state Republican convention, Steve Hilbert, a business tycoon with ties to Indianapolis and a long-standing friendship with the Trumps (he was part of a failed business deal to promote Melania Trump's makeup line), texted Pence to see if he would consider being Trump's running mate. Trump wanted to know if he would accept the offer if he was picked—he had already been rebuffed by Corker and Kasich, and it would be embarrassing if Trump offered it to him and it was rejected. Karen Pence, a ubiquitous presence, was huddled with her husband and Atterholt. To ensure privacy, Pence dismissed the state police guarding them and showed Atterholt the text and asked him what he thought. Pence said it would depend on what Trump envisioned, but he was clearly very interested.

In early July, the Pences went to Trump's golf club in Bedminster, New Jersey. After a round of golf, Pence, already keenly aware of Trump's desperate desire to be flattered, announced to the press that Trump "beat me like a drum." Privately, the Pences told the Trumps, "We've been praying for this meeting. We've been praying for *you*." The Trumps were stunned and replied, "What do you mean 'praying for *us*'?" The Pences told them they were "praying for wisdom, for clarity of thought." Neither Donald nor Melania knew what to say. According to a person familiar with that conversation, there was no cynicism or skepticism, but pure fascination. Karen and Mike talked about their Christian faith. "The Trumps were intrigued," the person said, "and that drew them closer together as couples." Three months later, Pence was touched when, on the night of the vice presidential debate, Trump left him a voice mail saying he had just said a prayer for him.

But when Trump called to offer Pence the job on July 14, he was not entirely enthusiastic. In a late-night phone call with senior aides, he sought their reassurance that he had made the right choice. He originally wanted to wait until the Republican National Convention to make the announcement, but Pence had to withdraw his name from his reelection race for governor by July 15, so the timing forced Trump's hand. On the day of the deadline, Pence's lawyer filed the paperwork withdrawing his name from the governor's race. Trump called Pence to offer him the job and asked him to pass the phone to Karen after they spoke because he knew she was by her husband's side.

When Dave McIntosh, a friend of Pence's, talks about Pence's decision to join the ticket, he compares Pence and Trump, rather grandiosely, to Jonathan and David from the Bible. The two were natural rivals for the crown but they became friends. "Typically," McIntosh said, "those two would hate each other because they're rivals, but they have this friendship and bond of brotherly love, and Jonathan's view was that he wanted to support David in what he needed. I think Mike's view is, *I might be in a place where I have differ-*

ent views, but my role now is to support the leader." Pence told McIntosh that if Trump asked him to join the ticket, he felt that it would be his calling and his mission in life to do it. For Pence, it is a sense of destiny, and he views it as part of a servant leadership model grounded in the Christian faith that extols the virtues of submission and obedience. The concept is rooted in the Bible, when Jesus washes his disciples' feet and says, "Whoever wants to become great among you must be your servant, and whoever wants to be first must be your slave." When Pence called his brother, Greg, to tell him that Trump had tapped him, they both cried. Greg, who is running for his brother's old congressional seat in eastern Indiana, told his little brother: "Well done, my good and faithful servant."

Pence's friends and colleagues speak in hushed tones about Trump's infamous 2005 *Access Hollywood* tape, afraid to say anything that might offend Pence's mercurial boss. In the tape, which was made public a month before the election, Trump was caught in a recording talking with host Billy Bush about sexually assaulting women. "When you're a star they let you do it," he said. "You can do anything . . . Grab them by the pussy. You can do anything."

Trump called Pence to apologize when the tape came out, but the Pences initially refused his calls. Eventually Pence answered and, after apologizing to him, Trump asked Pence to pass the phone to Karen so that he could apologize to her, too. She was particularly offended by his comments. It was a stunning development. "My brother would never say anything like that," said Greg Pence. "He's just not that kind of guy." As Trump's poll numbers took a beating, Pence, for a brief time, had the upper hand in their relationship. If he left the ticket, he could have torpedoed the campaign. The tape was made public on a Friday, and on Saturday Pence canceled a campaign appearance and said he could not defend Trump. He kept Trump waiting and did not express any public support until Monday, after spending a tense weekend deliberating with his wife and top aides. But the Pences stuck by Trump—their joint ambition overruling their concerns—and proved their unshakable loyalty. *Hate the sin, love the sinner* is an oft-repeated refrain that friends of the Pences use to describe their deliberations.

"I believe in forgiveness," Pence said, "and we are called to forgive as we have been forgiven."

IV
———————————

The Observatory

I always say, "People don't think you're human." And he just laughs.

—NAVY AIDE ON TALKING TO DICK CHENEY AT THE OBSERVATORY

L ocated at One Observatory Circle in Northwest Washington, about three miles from the White House and at the northwest end of Embassy Row, the Observatory is a hidden world with its own housekeepers, cooks, and traditions. Unlike the White House, the vice president's residence is not open for public tours. The elegant, white, nineteenth-century Queen Anne–style house sits on thirteen pristine acres where frustration simmers for some and power is cemented by others. The home used to be the official residence of the chief of naval operations, a four-star admiral at the helm of the navy. It is part of the U.S. Naval Observatory and is still run by the navy. The 9,150-square-foot, three-story house was built in 1893 and the grounds have been home to scientists who calculate the official mean distance between the Sun and Earth and who are considered authorities on astronomy and time calculation. A slew of atomic clocks and the largest collection of astronomical books in the United States can be found on the Observatory's expansive grounds. There are tours of the Observatory campus on Monday nights, when astronomers offer a glimpse of the heavens to visitors through an 1895-vintage twelve-inch refractor telescope.

In 1893 the Naval Observatory moved from a foggy spot on the banks of the Potomac River in downtown Washington to its current location on a hill, where the stars are much easier to view. It is a highly coveted and little-known property in Washington and it was the Observatory superintendent's residence before the chief of naval operations commandeered it in 1929 because it was so lovely. Congress passed a law in 1966 that set aside ten acres on the grounds of the Naval Observatory for an official home to be built for the vice president. Vice President Hubert Humphrey, looking ahead to his own presidential campaign in 1968, asked that the project be delayed because he worried that spending the seventy-five thousand dollars allocated to build the home would be considered a frivolous expense as the Vietnam War raged on. In July 1974, 174 years after the president first occupied the White House, Congress designated the already-occupied home as the official vice presidential residence. The resident at the time, Admiral Elmo Zumwalt, was so furious that he was expected to give up his home that he ran unsuccessfully against Virginia senator Harry Byrd, who was head of the congressional committee that approved the decision.

The charming thirty-three-room mansion has a wraparound porch, hunter green shutters, a white turret, and seven wood-burning fireplaces. It is known to White House staff as the VPR, "vice presidential residence," or NAVOBS, short for its more formal name, the Naval Observatory. The house and the scientific offices that surround it resemble a leafy New England college campus. Mature magnolia and boxwood trees are scattered throughout the grounds, making it feel more like a sanctuary than a "great white jail," as the White House has famously been called. Vice presidents have the best of both worlds, living in the Observatory and working at the White House. They can sit on their porch and no one will be standing at a fence gazing into their yard. It is much more personal and private than the White House. But, like most old homes, it has its quirks. Part of the basement floor was still covered in dirt when the first vice president moved in, hot water was not always reliable, and rust-colored water came out of the faucets.

George H. W. Bush preferred living in the Observatory, where he and his wife, Barbara, lived for eight years, over life at 1600 Pennsylvania Avenue, where they lived for four. Barbara said, "There was a big difference between opening the door at the Vice President's House, where there wasn't a soul around, in your bathrobe and letting the dog out at 6:00 a.m., and throwing on your warm-up suit at the White House, where the morning crew already was hard at work." Bush's own vice president, Dan Quayle, lived in the Observatory for four years and echoed the Bushes's feelings. Life in the White House, he said, is "lonely" and "very cold." At the White House, you walk outside and you're in downtown Washington; at the Naval Observatory there is a tennis court, a basketball court, and a swimming pool dotting the lush acreage. "You're away from everything," Quayle said. And the president and first lady rarely visit. Asked why not, Mondale said, with a smile, "*You* go to the president." But Secret Service agents are one part of the package no sitting vice president and his family can escape—they trail them even if they just want to take a walk around the grounds.

The home, with its large rooms and elegant furniture, feels more like a house from the set of *Gone with the Wind* than the place where Dick Cheney received classified briefings early each morning ahead of the president, or where lawmakers and donors have gathered for thousands of receptions over the years, or where vice presidents plan their own presidential campaigns. If the private residence of the White House, located on the second and third floors of the 132-room mansion, feels like an elegant Manhattan apartment—as it was described by Michelle Obama's first press secretary, Katie McCormick Lelyveld—then the Observatory feels like a country estate.

But there has been very little time for stargazing for the most recent occupants of the house.

———————

Gerald Ford was the first vice president eligible to live at the Observatory, but he assumed the presidency before renovations were completed. Ford's vice president, Nelson Rockefeller, an heir to the Rockefeller fortune, used the house mainly to entertain. He had a much more luxurious and even more private twenty-seven-acre estate on Foxhall Road that was one of the most expensive homes in Washington. Rockefeller spent only one night in the Observatory, but he made a striking contribution to its décor when he installed a mink-covered bed designed by famed German artist Max Ernst said to be worth $35,000. (It was later moved to a museum after a backlash over its exorbitant price tag.) The first full-time vice presidential occupants of the Observatory were Walter and Joan Mondale, who took up residence in 1977. Every vice president since has lived there.

Unlike the White House, where changes to the main floor have to go through an elaborate clearance process, redecorating is a much more casual affair at the Observatory, and each family has left its mark. The Quayles installed the heated pool, pool house, and putting green; the Gores removed the putting green and planted native trees and shrubs; the Cheneys borrowed art from museums around the country to decorate; and the Bidens have left behind seventy-five "Biden blue" Lenox china place settings with the vice presidential seal.

Like presidents and first ladies, vice presidents and their wives bring in interior designers to redecorate, and they make use of some of the furniture that is already there. Interior designers donate their services because it is considered an honor to work on the historic home. The Cheneys' social secretary, Elizabeth Haenle, who also managed the residence, created a museum database to catalog and inventory everything in the house with its provenance and valuation, and helped establish a secure storage facility to house pieces not being used, some dating back to the Rockefellers. (There is an intricate system in place to do the same for every piece in the

White House collection, which is stored in a climate-controlled facility outside of Washington. White House pieces date back to the eighteenth century.) The Vice President's Residence Foundation provides private funds for the redecoration.

"When we first moved into the residence, I would see navy enlisted aides who had no idea they were using a valuable pottery jar that Mrs. Mondale had commissioned Native American Indians to make for the residence as a milk creamer," said Haenle. "They had no idea because the continuity of the enlisted aides and the political staff wasn't there." The staff is composed of fewer than a dozen navy enlisted aides who cook, clean, and maintain the home—by comparison, the White House has about a hundred residence staffers, most of whom are not in the military. The longest a naval aide might work at the Observatory is two to four years because of career rotation in the Navy and eight years if they are a political appointee. Like at the White House, they work in shifts, with an early 6:00 a.m. shift and an evening shift that starts at 2:00 p.m. When the Cheneys left, they handed over the database and inventory to the Bidens. Haenle and her staff worked with Lynne Cheney to create a vice presidential library with books written about every vice president dating back to John Adams, and memoirs written by vice presidents and their wives.

The navy aides are usually experienced cooks who are loyal and discreet. Because it is such an intimate environment, friendships between the staff and the vice presidents they serve can be deep and long-lasting. The Cheneys brought a navy aide to work with them at their home in McLean, Virginia, after leaving office. The staff sees a side of the vice presidents that few others ever get to see. A person who worked for the Cheneys at the Observatory said that she always was a little intimidated by him. "He didn't have to demand respect, his voice commanded it. When he speaks, you listen." But his approval ratings during his eight years in office were often abysmal. She told him on more than one occasion, "People don't think you're human," which always made him laugh. Cheney said, "I don't feel sorry for myself or feel that I am unjustly or unduly criticized by those who disagree with me. I mean, I don't worry about it." But the aide could remember seeing Cheney in the summer of 2001, shortly after his family moved in and after he was fitted for a pacemaker (following his fourth heart attack at the end of 2000). She saw him sitting in the sunroom surrounded by mountains of flowers, and in that moment she saw how vulnerable he was. When Cheney needed his implanted defibrillator replaced in 2007, his doctors had the wireless feature disabled out of concern that a terrorist could hack into the device and shock his heart into cardiac arrest—a scene later portrayed in the Showtime drama *Homeland*.

Cheney's reputation among critics as a warmonger was a strange juxtaposition to how he behaved in the Observatory, where the Cheneys' grandchildren would stay for the occasional sleepover, putting their sleeping bags at the foot of their grandparents' bed. (Later, the four bedrooms on the third floor would become the favorite spot for sleepovers for the Biden grandchildren.) The Cheneys redid the upstairs kitchen because it had been a galley kitchen and they wanted more space. They also disliked using the formal dining room on the first floor; instead, when no official events were on the calendar, they ate their dinners on small trays while they watched TV or a movie upstairs. Cheney called the home "a very special place." It's a fantastic place for kids, he said, recalling how, whenever someone was having a birthday, they would have a big party on the lawn with donkeys and clowns. "The kids loved it," he said. "You miss it when you leave."

Haenle said that they used a portion of their official entertaining budget to host wounded warriors for BBQs and concerts at the Observatory. Haenle's brother was serving in the military in Iraq and Afghanistan when she was working for the Cheneys, and she was in the unique position of working in a house that was at the epicenter of decision-making about the war. "It was much more personal when my brother was in Mosul and there were days when they were voting or an attack had happened, or a raid was being prepared, when the vice president would call me over when he came home from the office to tell me, 'Listen, I wanted you to know that today they voted in Mosul and they had the highest turnout of anywhere. Your brother and his team really were amazing to get that out.'" On more than one occasion the vice president started a dinner he was hosting at the residence by calling Haenle into the dining room and telling his guests, "I want you all to know Liz and the story of what her brother and his troops did today in Mosul." Once he called Haenle from Air Force Two to tell her he met a wounded soldier in Houston who knew her brother and he wanted to give her the soldier's contact information. "He never failed to remember that I was one of those close personal connections to the war." It took a physical toll on the vice president. There were nights, a Naval aide said, when she took dinner up to the Cheneys and no one ate a thing. "He pushed it around a bit. It's war, it's ugly."

———

The navy still owns and maintains the house. A navy engineer is in charge of upkeep and maintenance of the house and grounds, and there are two gardeners and a couple of maintenance workers. The house was not always modern living: when the Quayles lived there they had window air-conditioning units on the second floor, and when the Cheneys added central air-conditioning they had to deal with brick on brick and no insulation. The windows were painted shut and they would often get jammed. "Come up with a sledgehammer!" Lynne Cheney jokingly bellowed to an aide from her room upstairs. The Cheneys were asked to delay moving into the house for six weeks after Bush's first inauguration because new wood floors needed to be installed on the first and second floors. The old floors in the nineteenth-century house had been refinished so many times that nails were dangerously sticking up.

An aide at the Observatory who worked for the Cheneys and

the Bidens said, "The Cheneys are very introverted, very quiet, and Joe Biden and Jill Biden are very extroverted. They both keep their friends close. It was amazing because you have these moments when they forget you're there and you watch them watch their grandchildren and you see that sparkle in their eyes and that smile when they don't have the weight of the world on their shoulders, where they can just be in the moment with their wife and their grandchildren." The day before the 2009 inauguration, the staff gathered with the Cheneys for what they called "the last latte," a reference to the Cheneys' love of Starbucks coffee. "You had to collect yourself because at noon the next day there was a new family," said a navy enlisted aide. "I was a Dell computer and now I was a Mac, I had to drop all my operating procedures."

The Bidens filled the house with art on loan from the National Gallery. "I wanted it to feel warm and comfortable," Jill said. "I didn't want people to walk through the front door and feel like they can't sit on the sofa." New York designer Victoria Hagan, who has been described by one home magazine as "the reigning queen of restrained elegance," donated her design services and helped the Bidens decorate. Hagan described the residence as a "very welcoming home and not pretentious." The Bidens switched up some of the more muted tones that the Cheneys had used. Light cream colors were replaced with sapphire blues and deep greens.

The Bidens personalized the house in unique ways, one of which included drawing attention to other, lesser-known occupants of the Observatory. Jill created the Family Heritage Garden of the Vice President with the names of everyone who has lived in the house, including their pets, etched onto pavers and placed around a fountain off the front lawn. She got the idea from the White House Children's Garden, which honors presidential children and grandchildren. She and Carlos Elizondo, the Bidens' social secretary and residence manager at the Observatory, contacted each vice presidential family to get a list of names. They placed a sculpture of their beloved German shepherd, Champ, alongside the fountain.

And in 2010, Biden surprised Jill with a plaque on a tall tree on the residence grounds with a special inscription: "Joe Loves Jill, Valentine's Day 2010."

When the Pences visited the Bidens in the Observatory, one of the Bidens' granddaughters met with one of the Pences' two daughters and told her "what it's like having Secret Service following you everywhere," Biden said. The Pences are making their own mark on the historic home. They have hung a plaque over the fireplace with a passage from the Bible: "'For I know the plans I have for you,' declares the Lord, 'plans to prosper you and not to harm you, plans to give you hope and a future.'"

The Second Lady

I don't think that passes the K Factor.

—MIKE PENCE'S CLOSE FRIEND ON THE EVER-PRESENT
INFLUENCE OF PENCE'S WIFE, KAREN

*The life of a second lady is very different from the life of a first
lady. It comes with all the trappings and none of the luxuries.*

—MARGUERITE SULLIVAN, AIDE TO MARILYN AND
DAN QUAYLE AND LYNNE CHENEY

An antique red phone has sat on Mike Pence's desk since he was a member of Congress. Only one person has the number, and when that person calls Pence stands at attention. "When that phone rings, everything stops," said Pete Seat, an official with the Indiana Republican Party. One aide jokingly compared Pence's red phone to the red phone that Police Commissioner Gordon used to call Batman in the 1960s television show. The phone, a Christmas gift from Karen, reminded everyone who visited Pence in his office of her influence and stayed with him even when he was governor of Indiana. While it is not in any of Pence's vice presidential offices—it could be a source of embarrassment for a man who some say relies too much on his wife—the couple have kept it in the Observatory, where it serves as a reminder of their shared history as political partners. Pence had such a small group of top advisers that aides describe it as a triangle, not a circle. When he was governor there were three people in that triangle: Mike, Karen, and Bill Smith, who was Pence's chief of staff in Congress and in the governor's office and is still a close adviser to Pence. Aides say the triangular relationship has been in place since Pence first ran for office in the late 1980s. Karen is a sounding board and a counselor to her husband. If a decision was made and abruptly changed, Pence's staff knew Karen had something to do with it. "On personnel decisions she would often interview folks," Atterholt said. "You could tell that they had discussed these issues the night before."

Karen is an unassuming Midwestern mother of three who has been wildly underestimated. Headlines about her include the seemingly frivolous decision to install beehives at the Naval Observatory and her line of towel charms that help identify whose towel is whose, a business she started as the governor's wife and is now defunct. The company's website tries to make the case for the bizarre tchotchkes: "I have had so many times where I was swimming at a friend's beach house, pool, or lake house, using their matching beautiful beach towels. Lo and behold, I would go in the water for a dip or up to the house for a beverage, and when I came back to my towel, it was gone! Someone else had grabbed my towel unknowingly . . . because all of the towels looked the same." She is a trained pilot, enjoys painting, and worked as an art teacher. Her signature issue as second lady has been art therapy, using art to help treat people suffering from traumatic brain injuries and other health issues.

While towel charms and beekeeping are seen as hobbies fit for a suburban housewife, being a top political adviser is not. But Karen attended meetings when Pence was governor and acted as his chief counselor. When he was not in the room people who worked for him talked in hushed tones about "the K Factor." K for Karen. Sometimes, when Pence would bring up an idea that they had not discussed before, and after he and Karen left the room, his staff would explain the new strategy: "That's *the K Factor*." Or when he rejected an idea they would say, "I don't think that passes *the K Factor*." Karen has read Stephen Covey's bestseller, *The 7 Habits of Highly Effective People*, more than once, and it shows. Most surprising changes of direction, staffers say, would come after Pence had time to discuss an issue with his wife.

As the first lady of Indiana, Karen's influence included staffing and policy, from tax cuts to education reform. Pence often told his staff that he would have to go home and "sleep on" decisions and "talk to Karen." "Karen is protective of him," said Pence's friend David McIntosh. "She's got very good antenna. She seeks out people to be working for him who are loyal. She has thought through a lot of the same issues that Mike has and she bucks him up on a lot of very conservative issues based on her faith." When Pence was governor, she played a role in making sure the personnel that surrounded her husband were both competent and loyal. "She would meet with the key people Mike wanted to surround himself with before they were hired, and I know Mike asked and valued her opinion on his key hires," said Atterholt.

She provided high-level strategic guidance to her husband as a member of Congress and as governor, and still does in the White House, where she has an office in the Eisenhower Executive Office Building, near where her husband's staff works. She sat in on at least one interview when Pence was putting together his staff. "No question in my mind," said one friend of the Pences, "he would not make any major, or perhaps even minor, decision without Karen." During the 2016 campaign, Pence introduced his wife as "the highest-ranking official in the state of Indiana." It was meant as a joke, but it is closer to the truth than most people in the audience would suspect.

Karen Sue Batten was born in Kansas in 1957 and was raised just north of Indianapolis, where she met her first husband, Steve Whitaker, in high school, and where she was valedictorian. Karen and Whitaker, a medical student, married in 1978 when she was just twenty-one, and they divorced not long after. "We were kids," Whitaker said. She graduated from Butler University in Indianapolis with bachelor's and master's degrees in elementary education, and in 1983 she met Pence when she was playing guitar at Mass at St. Thomas Aquinas, a Catholic church in Indianapolis. Pence approached her after Mass and learned that her sister was enrolled in the same law school he attended. "When I first met Mike Pence, it was love at first sight," Karen said in a television ad from Pence's gubernatorial campaign. "On our first date, we went skating at the Pepsi Coliseum at the state fairgrounds. We skated around for a little while, then he reached over and took my hand."

After dating for eight months she had a small gold cross engraved with the word "Yes," and slipped it into her purse so that she could show it to him when he proposed. A month later, Pence asked her to marry him while they were feeding ducks at a local canal. He hollowed out two loaves of bread, one held a ring box and another a small bottle of champagne. Pence still has the cross Karen carried with her while they were dating, but he does not wear it because he's worried he'll lose it.

After marrying in 1985, they struggled with infertility for six years and were on an adoption wait list before Karen became pregnant with their son Michael, who is in the U.S. Marine Corps. "By the time they called us with a possible child, we already knew that I was pregnant with Michael," Karen said. "And we just felt like it wasn't right for us to still be on that list of parents who wanted to be considered by the birth parents, and so we withdrew our name. Of course, our son has never forgiven us. He goes, 'Really, Mom, I could've had a brother! Really? What were you thinking?!' But we just felt like God had shown us He was going to bring us a family, and we needed to pull our name off." They later had two daughters: Charlotte and Audrey.

"All I ever wanted to be was to be a mom," Karen said. "I didn't care about fame or fortune, big house, fancy career, nice car—none of that has ever been important to me. I just wanted to be a mom.

And so my main thing was, how could God put this desire in my heart and not bring me kids?" But, it seems, she wanted much more. It was Karen, friends say, who was the biggest influence on their decision to leave the Catholic church—after they got married they became evangelical Christians. A Pence aide called her a "prayer warrior" who, in the loading dock in the garage of the Quicken Loans Arena in Cleveland, where the Republican National Convention was held, led Pence's small team of advisers in prayer. On the campaign she was omnipresent, and a month into his vice presidency she accompanied her husband on his first overseas trip to the Munich Security Conference.

Friends of the Pences like to tell a story to make clear just how close they are as a couple. At the governor's prayer breakfast in Indiana the governor is traditionally seated at the head table and spouses are seated at a reserved table in the front of the room. When the Pences arrived and a coordinator explained the setup to then-governor Pence, he was not happy. "This is where you're sitting," the staffer said, pointing to the head table, "and Karen will be down at table two, right in front." Pence said he needed his wife to sit with him, but he was told that could not happen. "That's OK," he said, "I'll just go down and sit with her then." After a few minutes of discussion with the event planners, the host returned and told Pence, "We'll rearrange things up there."

"It's the first time I've ever seen a first lady of Indiana have an office on the ground floor of the Capitol," said former Indiana attorney general Greg Zoeller, who has known the Pences for years. "I can't remember who got moved out, but the office of the attorney general is on the north atrium and just down the hall there's the court of appeals and right next to that was the first lady's office and she even had office hours. She was a player, not just as a consultant and an adviser to the governor, but she had an office eight feet down the hall from mine!"

Former Indiana secretary of state Ed Simcox has known Pence for more than thirty years. The two attended Bible study together when Pence was governor and occasionally, Simcox said, he emails Pence now with a word of encouragement or a Bible verse or quote. Karen is undeniably influential, he says. "I've known eight governors of Indiana. Five rather intimately. Karen is the ultimate example of a political wife who is a totally engaged member of the team, with opinions, with suggestions, with critiques." Mike Pence, he said, "is not complete without Karen."

Friends say Karen is more conservative than her husband, especially on social issues, and has backed Pence up when he was governor in his opposition to gay marriage and his support of the Religious Freedom Restoration Act, a law that makes it easier for businesses to discriminate against gays. In 1991, Karen wrote a letter to the editor in the *Indianapolis Star*, complaining that the paper's "Children's Express" section had run a story that "encourages children to think they're gay or lesbian if they have a close relationship with a child of the same sex" or are close to a teacher of the same gender. She wrote, "I only pray that most parents were able to intercept your article before their children were encouraged to call the Gay/Lesbian Youth Hotline, which encourages them to 'accept their homosexuality' instead of reassuring them that they are not." Pence's friends from Indiana say that the pair consider the world engaged in a moral conflict between those who are against Christianity and those who support it.

Bill Oesterle managed Republican Mitch Daniels's campaign for governor in 2004 and originally supported Pence, but says he worried that he was too easily influenced. "It's pretty clear that Pence hears voices," Oesterle said. "He pulled the federal funding for pre-K in the eleventh hour and no one knows what the influence was. Someone whispered to him that it was going to impact him nationally, that it was going to hurt him with Republican donors." Oesterle, who was the CEO of Angie's List until he stepped down in 2015, said he suspects that Karen played a role in Pence's about-face. In 2014 Pence rejected tens of millions of dollars in federal preschool funding, citing fears about "federal intrusion," but in

2016, during his tough fight for reelection before he was named Trump's running mate, Pence said he wanted to revisit the subject of federal funding. Karen was featured in a television ad. In it she said, "education is a priority for Mike." Critics saw it as pure political opportunism. Oesterle added, "He prays on stuff, and amazingly God is really looking out for his political interests."

The Pences live modestly—they rented a small colonial house in Washington, D.C., before moving into the Naval Observatory. (According to a campaign-finance disclosure form, in 2016 Mike Pence had one bank account, and it had less than fifteen thousand dollars in it—something that both alarmed and intrigued Trump.) Trump's inaugural committee raised an unprecedented $107 million, and while the group pledged to give leftover money to charity, eight months after the inauguration no money had yet gone to charity, and funds were used to help pay for the redecoration of the White House and the Observatory. That kind of spending runs contrary to the image Karen and her husband channel of 1950s Eisenhower Republicans. "When he got picked my first thought was, *Oh dear, how is Karen going to deal with Melania?*" said Scott Pelath, Democratic minority leader in the Indiana House of Representatives who worked with Pence. For her official White House portrait Melania wore her twenty-five-carat-diamond tenth-anniversary ring that reportedly cost $3 million (her engagement ring is a twelve-carat diamond that cost $1.5 million). Trump's wealth stands in stark contrast to the understated Karen, whose inaugural gown was made by seamstresses at a small Indianapolis shop called Something Wonderful; the store's original owner created her wedding dress and came out of retirement to help. (Karen had two dresses made and waited to find out what color Melania was wearing before she decided which to wear—the first lady's assistant told her she did not need to worry, she should wear what she wanted.) The Pences can be remarkably old-fashioned and stilted in their interactions with each other. When Pence was governor he invited Democratic members of the state legislature to the

governor's mansion for dinner. One attendee recalled Pence referring to Karen as "mother." "Mother, Mother," he said, "who
prepared our meal this evening?"

Far from the public image of a woman content to stay at home
and who "just wanted to be a mom," it is as though Karen has been
planning for this moment her entire life. On the piano in the living
room of the Observatory, there is a framed photograph of Karen
with Barbara Bush in the 1980s, when Bush hosted a lunch for the
spouses of Republicans running for Congress. Karen kept a paper
napkin from the visit with the vice presidential seal on it and placed
it in a scrapbook.

———

Karen Pence's predecessors Lady Bird Johnson, Pat Nixon, Betty
Ford, and Barbara Bush are all second ladies who went on to become remarkable first ladies. Life changed dramatically once they
moved into the White House. Craig Fuller, George H. W. Bush's
chief of staff when he was vice president, remembers how difficult
the relationship was between Nancy Reagan, as first lady, and Barbara Bush, as second lady. "Nancy Reagan was a very complicated
person who was fiercely loyal to Ronald Reagan," Fuller said, add-
ing that Nancy never appreciated the sacrifices Barbara made for
her. In some ways, he said, Nancy's obsession with her husband
prevented her from thinking about other people. The Bushes loved
entertaining, but they socialized less in the Observatory, Fuller
said, so they would not take any of the spotlight away from the
Reagans. Bush, he said, stepped back and did not articulate his
views on subjects to a group. "Barbara Bush didn't want the light
to be shining on her in a way that offended Nancy. I think it's fair
to say that Nancy never appreciated that."

In his personal diary, George H. W. Bush wrote in a 1988 entry,
"Nancy does not like Barbara." Nancy, he said, was jealous of his
wife. "She feels that Barbara has the very things that she, Nancy,
doesn't have, and that she'll never be in Barbara's class." The feel-

ing was mutual. When a negative biography of Nancy Reagan was published, Barbara snapped it up but slapped on another book jacket so that no one would know what she was reading.

Nancy and Barbara had such a frosty relationship that Barbara rarely set foot in the White House family quarters over those eight years that her husband was vice president. Paula Trivette was George H. W. Bush's nurse when he was vice president (as with presidents, vice presidents are taken care of by White House medical staff), and she, too, thought the underlying cause of Nancy's animosity was her jealousy. Nancy, Trivette said, was jealous of the close relationship Barbara had with her children and grandchildren. The Reagans had a famously strained relationship with their own children. "I think Mrs. Reagan wished in her heart that she had that type of relationship," Trivette said.

The loving marriage and deep mutual devotion that the Reagans shared was also true of the Bushes. Barbara defended her husband, who, as vice president, was dispatched to attend so many funerals for heads of state that he became a punch line for late-night television hosts. "George met with many current and future heads of state at the funerals he attended, enabling him to forge personal relationships that were important to President Reagan— and later, President Bush," she said. Barbara told Fuller during the early days of her husband's vice presidency: "I will do almost anything you ask me to do," yet added a very personal addendum: "But if at all possible, I'd like to end my day in the same city where George is."

Traditions that help define the job of second lady have eroded in recent years. Senate spouses have a bipartisan lunch on Tuesdays, and Senator Roy Blunt, a Republican from Missouri, said that his wife had to remind Karen Pence that she is technically the president of the Senate Spouses Club. Lady Bird, Pat, Betty, and Barbara all took the position as the presiding officer of the Senate Wives' Club (now the Senate Spouses' Club in acknowledgment of the growing number of women in Congress) very seriously. When George

H. W. Bush became vice president in 1981, making Barbara Bush president of the Senate Spouses' Club, she told the group: "I'm so happy to be here, I finally get to be a Senate wife!" (Bush ran unsuccessfully for the Senate twice, in 1964 and 1970.) Now, with so many members of Congress deciding not to move their families to Washington, the club is largely defunct, with fewer than a dozen members meeting once a month.

Happy Rockefeller, Joan Mondale, Marilyn Quayle, and Tipper Gore are modern second ladies who never became first lady. Nelson Rockefeller's wife, nicknamed "Happy" because of her sunny disposition, never lived in the Observatory, preferring instead to spend her time at their New York estate and their Washington mansion; she is best known for going public with her breast cancer diagnosis and the two mastectomies she had five weeks apart. She credited First Lady Betty Ford's brave announcement of her own breast cancer diagnosis a few weeks before for saving her life. It was Betty's experience that made Happy examine her own breasts. Happy was a socialite whose 1963 marriage to Nelson, the most famous scion of America's most notable wealthy family, was marred by rumors of their extramarital affair and the decision they both made to abandon their respective spouses and children. The marriage was said to have partly cost Rockefeller the Republican presidential nomination in 1964. Happy's successor, Joan Mondale, was nicknamed "Joan of Art" because of her work to promote the arts. She was a student of pottery, and when she moved into the Observatory she showcased contemporary art, pottery, and American glassware. She made her mark as a leader in the national crafts revival of the 1970s and she chaired the federal Arts and Humanities Council.

The partnership between Karen and Mike Pence is more closely aligned with the interdependent relationship between Hillary and Bill Clinton and Marilyn and Dan Quayle. At the 1988 vice presidential debate, sixty-seven-year-old Democrat Lloyd Bentsen famously humiliated forty-one-year-old Dan Quayle: "I served with

Jack Kennedy," he said, his tone escalating toward steady indignation. "I knew Jack Kennedy. Jack Kennedy was a friend of mine. Senator, you're no Jack Kennedy." Two aides to Quayle on the campaign said that Marilyn was the person who kept telling her husband to use the comparison, that his experience matched that of John Kennedy's when he ran in 1960. A friend of Quayle's said his aides were skeptical from the start about Marilyn's suggestion. "In politics you don't compare yourself to Kennedy. He has become mythical."

Quayle said his wife, whom he met at law school at the University of Indiana, was his de facto campaign manager during his first congressional campaign in 1976. Many of their classmates thought Marilyn was better suited to run for office than he was. "She has very good political instincts," he said. "I think [then Indiana governor Bob Orr] would have appointed her senator to fill my vacancy. When I became vice president Governor Orr and I talked about it. I talked to the president about it, and neither one of us thought it was a great idea. Bush said, *Whatever you want to work out*. But I know him well; it would not have worked. She would have been a great senator, she had many folks wanting her to run for governor in Indiana and later in Arizona."

Marilyn told the *Washington Post* that before the 1980s politicians never acknowledged that "your little wifey . . . helps you." Marilyn was her husband's fiercest defender and was unhappy with how he was being treated in the press. She became suspicious of White House aides, including James Baker, who questioned the wisdom of Bush's selection of Quayle. Like first ladies, second ladies undergo constant scrutiny, though to a much lesser degree because they are less visible. When she was second lady, Marilyn did not wear her wedding band for a photo shoot and there was immediate speculation that the Quayles were headed toward divorce when, in fact, she was late that morning and forgot to put her ring on after taking a shower. Quayle said his wife had "a very professional relationship" with First Lady Barbara Bush. But, he added, "It was not a mother/daughter relationship." The Quayles were decades younger than the Bushes and were raising their three children. Their children played with the Bushes' grandchildren in the White House and occasionally at Camp David, the president's rustic retreat nestled in the Catoctin Mountains about sixty miles outside of Washington.

Marguerite Sullivan, Marilyn's chief of staff, said the Quayles were treated unfairly by the media. Sullivan recalled Marilyn personally helping the mother of a student at her daughter's school who had breast cancer and was having trouble with her healthcare coverage. "A lot of them are portrayed as steel princesses, but they're not at all. Marilyn did a lot to help people, she herself would pick up the phone or get someone on her staff to make a call." There are challenges that come along with being second lady—apart from the name itself seeming old-fashioned and diminutive. "The life of a second lady is very different from the life of a first lady," Sullivan said. "It has all of the trappings and none of the luxuries." Unlike the first lady, the second lady often cannot use a government plane but still has to travel with a security detail. Second ladies often fly commercial or rely on chartered planes.

Like their husbands, second ladies must remember one principle: never outshine the first lady. Lynne Cheney was naturally a better public speaker than Laura Bush, but she was not sent out on the campaign trail often, and when she was it was to introduce her husband. Campaigns, Sullivan said, tend to send second ladies to smaller, less politically important cities, and because most campaigns are preoccupied with media markets they would not send a second lady to a bigger market than a first lady.

———

Al Gore stole his classmate's prom date. At a party after the 1965 prom at St. Albans, the elite all-boys prep school in Northwest Washington that Gore attended, Gore was introduced to a fun-loving girl named Mary Elizabeth Aitcheson. Her nickname,

"Tipper," was given to her by her mother and was from one of her favorite songs from her childhood, "Tippy Tippy Tin." She was someone else's date that night, but when she was introduced to Gore she thought, *Oh, boy! He's good-looking.* "We had a good conversation. We connected," she said. Wasting no time, Gore got her phone number and asked her out the next weekend. That summer he worked in Arlington, Virginia, not far from the Aitchesons' house, and stopped by to see her during his lunch break almost every day. Tipper made him bologna and cheese sandwiches. "Can't you make anything else?" he eventually asked her.

"No," she said simply. They got married in 1970.

When Gore told Tipper that he was running for Congress in 1976, she said, "I wanted to faint. They would have to bring me back with smelling salts." Tipper had a job in the photography department of the *Tennessean,* and when they were married Gore promised a quiet life in the country where he would buy a local paper and write books. But Gore was the son of a U.S. senator and from birth was destined for a life in politics, whether his wife wanted it or not. The Gores had four children and Tipper supported her husband's career, but she treasured her time alone and her privacy. She is artistic and was once the drummer of an all-female rock group called the Wildcats. She has played drums with members of the Grateful Dead and Willie Nelson. When Gore was in Congress, she built a darkroom in the basement of their home in Arlington, and has published books on photography.

And she used her husband's influence to promote causes she cared about. In the 1980s Tipper co-founded the Parents Music Resource Center after being horrified by lyrics on her daughter Karenna's Prince album *Purple Rain.* Her very public advocacy led to "Parental Advisory" warning labels on records with explicit lyrics. It makes sense, then, that Tipper was particularly disgusted by Bill Clinton's more-than-one-year-long affair with twenty-two-year-old White House intern Monica Lewinsky. Lewinsky was just a few years older than the Clintons' daughter, Chelsea, and the same age as the Gores' eldest daughter, Karenna. "She felt personally offended by the scandal," said Jamal Simmons, who worked on Gore's presidential campaign. A close aide to Gore in the White House said, "In Gore's case there's no way he would have gotten to such a dark place if Tipper hadn't been so mad." That "dark place" was Gore's decision to distance himself from Bill Clinton during the 2000 campaign, even as Clinton repeatedly offered to campaign on his behalf.

In the last two years of the Clinton administration, Tipper went to few White House events and, aides say, kept her distance from the president. According to Gore's vice presidential records, there were just a handful of phone calls and meetings between him and Bill and Hillary Clinton during the fall and early winter of 2000, after the November election and before the Supreme Court 5–4 decision that ended the recount of ballots in Florida and essentially called the election for Bush. Clinton and Gore had one two-minute conversation on December 13, 2000, a day after the devastating decision that ended Gore's presidential ambitions. According to Clinton's presidential daily diary, between November 6, 2000, and December 14, 2000, two days after the Supreme Court ruling, Gore and Clinton only spoke seven times with most calls lasting just a few minutes.

It is difficult in politics to separate what is business from what is personal, and elected officials have to do things all the time that are strictly business but feel personal, including firing staffers who have been loyal to them but who are no longer needed. It is different for family members, an aide said. "For them, it's all personal and if their families get slighted in some ways—which happens all the time—then for the politicians it's personal." When a late-night talk-show host made a suggestive joke about Gore taking his daughters through the gym locker room in the Senate, and imagined how excited Senator Ted Kennedy would be to see them, Tipper, for the duration of the campaign, banned Gore from going on the show again.

It was a far cry from the image forged in the early days of the 1992 campaign of baby boomers, the Clintons and the Gores, on a double date. When the two telegenic young couples set off on bus tours in the summer of 1992, they hit every major media market, and they were picture perfect. By all accounts, Al Gore and Bill Clinton were genuinely fond of each other and had more in common than their age—Clinton was forty-five and Gore was forty-four—and their Southern roots. Tipper and Bill were the fun-loving personalities, and Hillary and Al were more studious and reserved. (Tipper and Bill share the same birthday, except she is two years younger.) They made each other laugh, and a photo of Tipper and Hillary sitting on their respective husbands' laps on board the bus gave the impression of genuine affection between them.

In the early days of the administration, Tipper played a valuable role and became an advocate for people suffering from mental illness. "My work on mental health while in the White House is among the most gratifying things I've done in my life," she said. She is most proud of her role passing the Mental Health Parity Act in 1996. The bill required health insurers to cover mental health treatment. She held dinners in the Observatory with members of Congress from both parties to discuss mental health legislation.

Second ladies can operate under the radar in a way first ladies cannot. Tipper said that because she "didn't have the same level of media coverage as others in the administration," she was able to host roundtable discussions with teenagers around the country and hold private conversations with people struggling with mental illness. She invited Betty Ford and Rosalynn Carter, both of whom had been mental health advocates, for dinner at the Naval Observatory to get their ideas. "I believe it was I who suggested to President Clinton that we organize the first White House Conference on Mental Health," she said, "and he agreed." She still remembers a flight attendant who thanked her for her work on the issue. "Someone slipped a simple note to me once that said I had helped save his life. I still have that note."

After her husband ascended to the vice presidency she yearned for a connection to reality, and in Washington, with no fanfare, she walked through local parks and talked with homeless people, some of them mentally ill. She tried to persuade them to go to a shelter for a meal and a shower. "It all started one day when I was driving the kids home from lunch on the Hill with their dad. After passing a woman on the street, the girls asked, 'Why is that woman talking to herself?'" When Tipper replied, "Not everyone has a home, and many have a mental illness," her daughters asked, "Can we take her home with us?"

"We don't have room at home for her, but let's find a way to help," Tipper told them. Shortly after that, she and her children began volunteering at Martha's Table, a soup kitchen in Washington. She'd begun working with Health Care for the Homeless during the 1990s, and in 1994 she started driving around the Washington area with a friend in search of people with schizophrenia, depression, bipolar disorder, and other mental illnesses to try to convince them to go to a shelter or help them get treatment. She became friends with some of them, including a man named Captain Kersh who was a Vietnam veteran, and invited them for lunch at the Observatory. "They had become my friends and were leading productive, healthy lives eventually," she said. "So I was delighted to have them over to lunch at the Naval Observatory, and they were delighted to join me."

At the Observatory they toasted with sparkling cider. Some of her guests had no idea who she was. One guest told a friend of Tipper's, "Nice house. Her husband must have a good job."

"Well, he does," Tipper's friend said. "He's the vice president."

Trooper Sanders was a White House policy adviser to the Gores and remembered when there was opposition in the West Wing to a more aggressive approach to mental healthcare. Clinton took one of Tipper's aides aside after a meeting on healthcare and asked her, "What does Tipper want me to do?" "You have to be judicious when you're in the vice president's orbit," Sanders said. "You have to pick your battles. So when Mrs. Gore's pushing on something, because it doesn't happen every day, there's a lot of judgment and thought that goes into that." And it helps when one of the four principals—either the president or the vice president or the first lady or the second lady—has a stake in a particular issue, Sanders added. "Because the president knew this was something that Mrs. Gore cared about, mental health got more time and perhaps a greater look than might have been the case otherwise."

When Gore ran for president in 2000, Tipper revealed that she herself had been treated for depression. It was brought on by trauma she suffered after their son, Albert III, was hit by a car when he was six years old. He survived, and in her book *Picture This, A Visual Diary* she described the terror she felt. "I watched in horror as he flew through the air, scraped along the pavement and then lay still." (In his acceptance speech at the 1992 Democratic National Convention Gore was criticized for using the accident for political ends when he rather incongruously described "waiting for a second breath of life" from his son, and linking it to the nation which was "lying in the gutter, waiting for us to give it a second breath of life.")

Tipper received her bachelor's and master's degrees in psychology, and at one point she planned on becoming a therapist. "But," she said, "life went in another direction for me." Tipper's mother also suffered from depression and was hospitalized at least twice. "Once, when my mother was in the hospital for something else, I wanted to tell the doctors about the medication she was taking," Tipper wrote in her memoir. "But she was so terribly fearful of anyone finding out—even a doctor—that she wouldn't let me tell them. It broke my heart."

The famously affectionate Gores shocked everyone when they announced their separation in 2010, a couple of weeks short of their fortieth wedding anniversary. At the 2000 Democratic National Convention in Los Angeles the two shared an awkward but passionate kiss. After the devastating loss of the 2000 election, they began to drift apart and aides speculate that the family's history of depression made the loss even more difficult. "She was like, 'You don't want to stop working, you don't want to stop this. I don't want to do it anymore,'" said Sarah Bianchi, who worked for Gore and is close to the Gores' daughter Karenna. "I think it was very genuine. They really wanted different things." Bianchi said that for Karenna her parents' divorce was harder than her own divorce. Like the rest of the country Karenna had a vision of her parents as a team and hopelessly in love.

———

By the time Lynne Cheney moved into the White House she was used to being married to a politician. Dick Cheney had worked for Nixon and Ford, he was in Congress from 1979 to 1989, and he was U.S. defense secretary under George H. W. Bush. "The Cheneys are a very close family and politics is their family business," said Neil Patel, who was a policy adviser to Dick Cheney in the White House. "They all weigh in on everything freely."

Dick and Lynne both grew up in Casper, Wyoming, where Lynne's mother was the deputy sheriff. They met when they were students at Natrona County High School. Cheney was a popular football player and Lynne was a majorette, a dancer with a baton who is usually part of a school marching band. During one alarmingly dangerous routine, the ends of her baton were set on fire and Dick Cheney waited with a can of water to extinguish it as she twirled toward him.

They were married in 1964 and have two daughters. Their eldest daughter, Liz, is a Republican congresswoman from Wyoming, and their other daughter, Mary, is openly gay, a gay-rights activist, and one of her father's closest confidantes. The Cheneys' private life became public when, in 2004, John Kerry invoked Mary in a presidential debate when he was asked whether he thought being gay was a choice. "I think if you were to ask Dick Cheney's daughter, who is a lesbian, she would tell you that she's being who she was, she's being who she was born as." In the audience Laura Bush and her daughters, Jenna and Barbara, gasped. Lynne called it "cheap and tawdry." George W. Bush had come out in support of a constitutional amendment banning same-sex marriage, complicating his relationship with Cheney. "My general view is that freedom means freedom for everyone," Cheney has said. He thinks that Bush "agonized over [his position against gay marriage] more than I did."

Lynne is more self-assured and more outspoken than the reserved Laura Bush. She even considered running for the Senate from Wyoming in 1994. "I don't think they're buds," said a close Cheney friend of the relationship between Lynne and Laura. But the two women shared the painful experience of September 11 together. When Laura was hustled into the Presidential Emergency Operations Center, known as the PEOC, an emergency command center located in the basement of the White House, Lynne was one of the first people she saw. (Cheney, National Security Adviser Condoleezza Rice, counselor to the president Karen Hughes, and deputy chief of staff Josh Bolten were also there.) Lynne had been there since that morning. She hugged the first lady and whispered in her ear, "The plane that hit the Pentagon circled the White House first." The terrorists had originally planned to hit the White House but they decided on the World Trade Center, the Pentagon, and the U.S. Capitol instead.

A couple of days after 9/11, Lynne gathered her small staff, many of whom were women in their late twenties and early thirties, and they sat together on the beautiful veranda at the vice president's residence. She brought a copy of David Brinkley's *Washington Goes to War*, a book about the capital at the beginning of World War II. "She wanted us to talk about it," meaning 9/11, said the Cheneys' social secretary Liz Haenle. "What does this mean for us as a country, what did this mean for us as an administration who suddenly found ourselves at war." What followed were many nights when the Cheneys were told to leave their home and were brought to a series of "undisclosed secure locations" to preserve presidential succession in case another attack happened. Often the location they were brought to was no more exotic than Camp David, but it was done simply to keep Bush and Cheney apart in case something happened to Bush. Concerns about a "second wave" of attacks in the weeks and months following 9/11, and the anxiety of those ensuing days, is difficult to overstate. A color-coded schedule was devised to show Cheney's location and the president's so they would not accidentally be in the same place at the same time; Cheney often appeared in meetings via teleconference after the attack. Haenle said, "Continuity for the Cheneys was living out of a duffle bag."

Lynn has a doctorate in nineteenth-century British literature from the University of Wisconsin and is a bestselling author and college professor. She oversaw the National Endowment for the Humanities during the Reagan and George H. W. Bush administrations from 1986 to 1993, and, after she left government, she was a regular on CNN's *Crossfire,* as an unapologetically conservative woman who fought against political correctness. She has ardently defended the study of Western civilization while arguing it is being trumped by an emphasis on multiculturalism. Her occasional sign-off, until she stopped going on *Crossfire* in 1998 was, "From the right, and right on every issue, I'm Lynne Cheney."

She sounds more like Hillary Clinton, though with very different political views, than Laura Bush. She was an education adviser in 2000 to then-candidate George W. Bush, and her name was floated as a possible education secretary. Liberal commentator Bill Press, who often argued the opposing side in televised debates with Lynne, described her as a "fierce competitor" and "a true believer." During the 2004 election he joked, "John Edwards [who was John Kerry's running mate in 2004] should feel lucky that he's debating Dick Cheney rather than Lynne Cheney."

A *New York Times* profile written shortly after her husband became vice president described her as "tart, take-charge and too busy to shop." When her husband became vice president she was on the boards of several companies and resigned from two of them— Lockheed Martin Corp. and Exide Corp. Dick Cheney still sounds exasperated when he recalls the changes she had to make when he was elected so there would be no allegations of conflicts of interest. "It changed her world," Cheney said. Lynne said if a woman were president or vice president, "and that is going to happen," and the spouse was a man, "everyone would think it odd if he didn't continue with his career." She did not give up her job as senior fellow at the American Enterprise Institute, however, and continued writing about education when her husband was vice president.

After 9/11 Lynne found a new calling and worked on several American history books for children from a second-floor office at the Observatory. Several of her books, including *A is for Abigail* and *America: A Patriotic Primer*, were bestsellers, but a lesser-known work is a novel, *Sisters*, that was published in 1981 and features a lesbian love affair. Her biography on the White House website highlighted the nine books she wrote and notably excluded *Sisters*. She convinced her publishers not to reissue the book in 2004 during the reelection campaign. The book is set in the nineteenth-century American West, "when men were men, and women were property," and includes marital rape. Lynne had to address controversy surrounding the book in 2004 after Republican senator George Allen of Virginia used passages from controversial novels written by his Democratic challenger, Jim Webb, during the campaign. Webb said, "I mean we can go and read Lynne Cheney's lesbian love scenes if you want to, you know, get graphic on stuff." Lynne said, "I'm not going to analyze a novel I wrote a long time ago. I don't remember the plot." Today, a new copy on Amazon costs more than seventy dollars.

Sisters was surprising because of Lynne's politics, which are more conservative than her husband's. "She is as hardline conservative as Dick Cheney or more," said a close friend of the Cheneys'. "And she has a tough personality. She's hard-charging, he's very easygoing on a personal level. Cheney scares people when they hear his name but when they meet him, he's a nice guy." Lynne Cheney is more intimidating in person than her husband. The Cheneys' friend said, "When people meet her, they're like, 'Wow, I wasn't expecting that.'"

———

Jill Biden was nicknamed "Captain of the Vice Squad" inside the Obama White House. She dressed as a server at a party she threw, and hid in the overhead compartment of Air Force Two to scare Biden's top staffers during a trip. She has a doctorate in education and continued working at her job as an English professor after her husband became vice president—she is the only second lady to continue working full-time. She commuted from the Observatory to Northern Virginia Community College nearly every day during the week. She told Biden that she had to continue teaching after he became vice president because the job is part of her identity. She graded papers on Air Force Two, and friends say Jill was more drawn to teaching than to the job of being second lady. But over time the spotlight bothered her less. Though less involved than Karen Pence, Jill said that she sometimes Scotch-taped messages on Biden's bathroom mirror with an article that she wanted him to see, especially if it helped bolster her side of a policy discussion.

Biden is a Blue Star mom—Beau Biden was a captain in the Army National Guard—and she worked with First Lady Michelle Obama to honor military families. "She just seems to walk this Earth so lightly, spreads her joy so freely. And she reminds us that although we're in a serious business, we don't have to take ourselves too seriously," Barack Obama has said of Jill.

Trooper Sanders, who worked for Michelle Obama in the East Wing after he worked for Tipper Gore, said Michelle's and Jill's staffs coordinated seamlessly. On the military families initiative it was Jill who brought "authenticity" and "an expertise" to the issue. "It's great for a first lady to have such a strong partner in the second lady . . . because they can pass the baton back and forth when different things come up." Because Michelle Obama was a global celebrity, she realized that she could help Jill get attention for causes she cared about, like supporting community colleges.

When asked why she fell in love with her husband, Jill is quick to say, "I fell in love with the boys [Biden's sons, Beau and Hunter] first." Three years after his first wife, Neilia Biden, and their one-year-old daughter, Naomi, were killed in a car accident, Biden's brother set him up on a blind date with Jill Jacobs when he was thirty-three and Jill was a senior studying English at the University of Delaware. She worked part-time as a model and Biden recognized her from an ad. "I was a senior, and I had been dating guys in jeans and clogs and T-shirts, he came to the door and he had a sport coat and loafers, and I thought, 'God, this is never going to work, not in a million years.' He is nine years older than I am," she recalled. "But we went out to see *A Man and a Woman* at the movie theater in Philadelphia, and we really hit it off. When we came home . . . he shook my hand good night . . . I went upstairs and called my mother at 1:00 a.m. and said, 'Mom, I finally met a gentleman.'" After their second date, Biden asked her if she would mind not seeing anyone else. Biden proposed to Jill five times before she accepted. She wanted to wait because she wanted to make sure their marriage could work since she had grown so close to Biden's two young sons. "They had lost their mom," she said, "and I couldn't have them lose another mother. So I had to be one-hundred percent sure." Biden's sons were certainly sure about Jill. Beau told his father, "Daddy, we were talking and we think we should marry Jill," referring to a conversation he had with his younger brother, Hunter. She finally accepted in 1977 after Biden gave her an ultimatum.

The Bidens' social secretary, Carlos Elizondo, describes the Bidens as "very social" and talks about Jill as a friend and not a boss. On Inauguration Day in 2009 he told Jill he would move them into the Observatory and then he planned to attend an inaugural ball that night. He went home to get ready and was surprised to get a call from Jill on her cell. She was getting her hair done ahead of the balls and checking in to see how he was doing. He could not believe that she would be thinking of him at such an important moment in her life. When Elizondo's mother passed away the Bidens hosted a private memorial Mass and brunch at the Observatory for about a hundred of his family and friends. "That is who they are and I will never forget it."

In the 2008 election campaign, Joe Biden's description of his wife as "drop-dead gorgeous" bothered some Democratic voters. Jill, however, was not bothered in the least. "Sometimes I get a little put off by things he might say that are too personal for me," she said. "But, the thing is, I think Joe believes that." She laughed. "How can you get offended when your husband thinks that about you?"

VI

Tragedy and Trauma

I've lived a week since this morning.
—LYNDON B. JOHNSON ON THE FLIGHT BACK FROM DALLAS, NOVEMBER 22, 1963

Since World War II the vice presidency has become strikingly important. Five twentieth-century vice presidents have succeeded to the presidency because of tragedy, and, in one case, self-destruction: Theodore Roosevelt when William McKinley was killed in 1901; Calvin Coolidge, when Warren G. Harding died in 1923; Harry Truman, when FDR died in 1945; Lyndon B. Johnson, when John Kennedy was assassinated in 1963; and Gerald Ford, when Richard Nixon resigned in 1974.

On March 1, 1945, in the final months of World War II and not long after being elected to an unprecedented fourth term, President Franklin Delano Roosevelt addressed a joint meeting of Congress seated, an unusual position for the commander in chief. He was unable to bear the pain of standing with the heavy braces he wore after contracting polio, a paralyzing illness, at age thirty-nine. "Mr. Vice President, Mr. Speaker, and members of the Congress," he said, "I hope that you will pardon me for the unusual posture of sitting down during the presentation of what I want to say, but I know that you will realize it makes it a lot easier for me in not having to carry about ten pounds of steel around on the bottom of my legs, and also because of the fact that I have just completed a 14,000-mile trip." Roosevelt had just returned from the Yalta conference in Crimea, where he laid out plans for Europe's postwar reorganization with British prime minister Winston Churchill and Soviet premier Joseph Stalin. "As soon as I can," Roosevelt told his vice president, Harry Truman, after his address, "I will go to Warm Springs [a warm-water resort in Georgia that helped ease his pain] for a rest, I can be in trim again if I can stay there for two or three weeks." Roosevelt left Washington on March 30, 1945, and as Truman hauntingly wrote in his memoir, he "never saw or spoke with him again."

On April 12, President Franklin Roosevelt died at his cottage in Warm Springs, with his longtime mistress, Lucy Mercer Rutherfurd, by his side. His wife, Eleanor, was at work back in Washington; she had delivered a speech that afternoon and was listening to a piano performance when she was told to return to the White House. When she learned of her husband's death she immediately called their four sons, who were on active military duty, and changed into a black mourning dress. Just before 5:00 p.m., Roosevelt's vice president, Harry Truman, was on Capitol Hill in House Speaker Sam Rayburn's office. As he walked in, Rayburn told Truman that Roosevelt's press secretary, Steve Early, had just called looking for him. When Truman reached Early his voice sounded unusually strained. "Please come right over," he told Truman, "and come in through the main Pennsylvania Avenue entrance." Truman asked Rayburn to keep the call quiet, assuming the president had cut short his trip to Warm Springs to go to the funeral of his friend Bishop Atwood. It was odd, though, as on the rare occasion when Roosevelt summoned him for a secret meeting he asked him to use the east entrance so that no one would see him. Because the president was out of town, Truman had more Secret Service agents protecting him than usual and he longed for privacy. So instead of leaving Rayburn's office and going to his own office in the Capitol, where his agents were waiting to bring him to the

White House, Truman ran through the basement trying to escape his stifling security detail. "This was the only time in eight years that I enjoyed the luxury of privacy by escaping from the ever-present vigil of official protection," Truman wrote in his diary. He had no idea that he was about to become president and have a far bigger security detail.

Around 5:30 p.m. Truman was at the White House and was brought up to Eleanor's study in the residence on the second floor, where the first lady, Early, Anna Roosevelt Boettiger, the Roosevelts' only daughter, and her husband, Colonel John Boettiger, were gathered. "Harry," Eleanor said softly, placing her hand on Truman's shoulder, "the president is dead." Truman, in shock, let a moment pass and asked her if there was anything he could do to help. She shook her head and asked, "Is there anything *we* can do for *you*? For you are the one in trouble now." Secretary of State Edward Stettinius knocked on Eleanor's door. He was in tears. Truman called a meeting of the Cabinet, arranged for Eleanor to go to Warm Springs, and got a car for his wife, Bess, and daughter, Margaret, to bring them to the White House. He called Chief Justice Harlan Fiske Stone and asked him to come to deliver the oath of office. After scrambling to find a Bible, with his wife and only daughter standing inches to his left (with the exception of Secretary of Labor Frances Perkins, they were the only women there) and a scattered group of Cabinet members and members of Congress surrounding them, the chief justice raised his right hand and Truman held the Bible in his left and placed his right hand on top of it. Truman placed an index card on the Bible with the oath printed on it so that he wouldn't stumble on any of the words. "I, Harry S. Truman, do solemnly swear that I will faithfully execute the office of the President of the United States, and will to the best of my ability, preserve, protect and defend the Constitution of the United States." A little after 7:00 p.m., after a ceremony that took less than a minute, and less than two hours after he learned of FDR's death, Truman was now in charge of one of the deadliest wars in world history.

When Roosevelt was alive, Truman tried not to think about his health and what was essentially Truman's entire reason for being—to succeed to the presidency in case something happened to Roosevelt. The president only indicated once that he was not well when he asked Truman, in the fall of 1944 when Truman was campaigning, how he was planning to travel. When Truman said by plane, Roosevelt vetoed the idea: "One of us has to stay alive." Like many of his predecessors, Truman, who was vice president only from January 20 until April 12, 1945, was not terribly happy in the subservient role. He famously said, "The vice president simply presides over the Senate and sits around hoping for a funeral." He wrote to his daughter, Margaret, and said, "Hope I can dodge it [the vice presidency]. 1600 Pennsylvania Avenue is a nice address but I'd rather not move in through the back door—or any other door at 60." Shortly before being nominated as FDR's VP, Truman told a reporter: "Do you recall what happened to most Vice Presidents who succeeded to the Presidency? Usually, they were ridiculed in office, had their hearts broken, lost any vestige of respect they had had before. I don't want that to happen to me." After Roosevelt's death, Truman met with the Cabinet and asked its members to stay on. Roosevelt was beloved, and Truman promised to continue his policies and needed their help to do so.

When the short meeting was over and members of the Cabinet left the room, one person stayed behind: Secretary of War Henry Stimson. He had an important message to deliver to the new president. An enormous project was under way, "a new explosive of almost unbelievable destructive power." Truman recalled how "puzzled" he was by this and by the fact that Stimson felt he could not say more at that moment. "I had known," Truman wrote, "and probably others had, that something that was unusually important was brewing in our war plants." In the Senate, as chair of the Committee to Investigate the National Defense Program, Truman had sent investigators to factories where weapons were being manufactured to find out what was happening. Stimson then had paid him a visit. "Senator," he said, "I can't tell you what it is, but it is the greatest project in the history of the world. It is most top secret, many of the people who are actually engaged in the work have no idea what it is, and we who do would appreciate your not going into those plants." Truman complied, ended the investigations, and never heard another word of it until the day Roosevelt died.

The day after that Cabinet meeting, Truman learned more from Jimmy Byrnes, who was Roosevelt's director of war mobilization. The United States was "perfecting an explosive great enough to destroy the whole world," Byrnes told him. Later, Vannevar Bush, who was head of the Office of Scientific Research and Development, visited Truman at the White House. "That is the biggest fool thing we have ever done," he said. "The bomb will never go off, and I speak as an expert in explosives." But Truman soon learned that Bush was wrong when the bomb was successfully tested on July 16 in New Mexico. Truman had only met with FDR twice in private after they were inaugurated. Before becoming president, Truman did not know about the $2 billion Manhattan Project that resulted in the creation of the atomic bomb, and he found himself faced with the most terrible decision any president has had to make: to end World War II by dropping atomic bombs on Japan. Less than four months after assuming the presidency, on the morning of August 6, 1945, an American B-29 bomber, the *Enola Gay*, dropped the world's first atom bomb over Hiroshima. Bombs dropped on Hiroshima and Nagasaki killed more than 200,000 people.

In his February 5, 1953, diary entry, after eight years as president, Truman called the presidency "the greatest office in the history of the world—the greatest honor and the most awful responsibility to come to any man." A letter to his wife, Bess, who often escaped the White House to return to the peace of their home in Independence, Missouri, dated June 3, 1945, makes his isolation clear: "Dear Bess," he wrote. "This is a lonesome place." Now he had the top job and was astonished by the isolation that comes with it.

The deeply private Bess had been second lady and first lady, and she much preferred the less visible role of the former because it allowed her and the Trumans' daughter, Margaret, to lead a fairly normal life at their apartment at 4701 Connecticut Avenue, less than five miles from the White House. While thousands of Harry Truman's notes and letters are publicly available at his presidential library in Independence, Missouri, Bess burned most of hers. "One evening in 1955, around Christmastime, Grandpa came home and found her in front of the fireplace with a roaring fire, throwing in bundles of her letters to him," said Truman's eldest grandson, Clifton Truman Daniel. "And he stopped her and said, 'Bess! Dear God, what are you doing, think of history!' And she said, 'Oh, I have,' and kept throwing letters in the fire. She destroyed almost all of them."

Truman's ascension to the presidency at such a consequential time in American history has made him one of the most important presidents of the twentieth century. Just before leaving the White House, Truman hosted a small dinner for British prime minister Winston Churchill. Churchill made a confession to Truman during his visit to Washington: he was concerned when Roosevelt died and Truman had to take over so suddenly at such a perilous time. "I misjudged you badly," Churchill said. "Since that time, you, more than any other man, have saved Western Civilization."

In many ways fate determines the vice president's role and whether he will be remembered as a key figure in history or as an occasional stand-in for the president. "The main role of the vice president is to take over when something happens to the president," Cheney mused. "Harry Truman and others were there to do what the job was created for . . . Teddy Roosevelt—we got some great presidents because something happened."

———

"I'm forty-three years old," John F. Kennedy told his aide Ken O'Donnell, to calm his nerves after Kennedy named Lyndon Johnson as his running mate, "and I'm the healthiest candidate for president in the United States. You've traveled with me enough to know that I'm not going to die in office. So the vice presidency doesn't mean anything." Sometimes Kennedy would joke about his own death, like when he knew he would have to fly through bad weather on Air Force One on one particular trip to Ohio. "If that plane goes down, Lyndon will have this place cleared out from stem to stern in twenty-four hours," Kennedy joked to his friend and aide Ted Sorensen, with his valet George Thomas listening in as he buttoned up his shirt. "And you and George will be the first to go." Just when you think the position doesn't mean something, it does. Less than three years later Vice President Johnson would become president after Kennedy's assassination.

Concerns that Johnson would be dumped from the ticket in the reelection year of 1964 were real. Johnson had spent ten hours and nineteen minutes alone with Kennedy in their first year in office, but by the third year they were down to one hour and fifty-three minutes. At the president's surprise forty-sixth birthday party on May 29, 1963, Johnson was absent—no one had remembered to invite him. On the Hill, Johnson was furious, devastated by his impotence. "I'd like to get out of this damn town, go back to Texas and never come back," he yelled. Kennedy's personal secretary, Evelyn Lincoln, recorded a startling conversation in her diary before Kennedy left for his fateful November 1963 trip to Texas. Kennedy sat in the rocking chair in the Oval Office, with his legs crossed and his head resting on the back of the chair, and mentioned that naming a running mate could wait until the convention. Thinking out loud, but according to Lincoln without hesitation, the president said, "At this time I am thinking about Governor Terry Sanford of North Carolina. But it will not be Lyndon." One of Johnson's longtime aides, Bobby Baker, was being investigated by the Senate for bribery and Johnson was getting embroiled in the scandal. *Life* magazine was investigating the vice president's personal finances and editors were discussing whether to run a story in the next issue.

Johnson was aware that he was in danger of losing his job—no matter how much he hated the vice presidency, being forced out of it would have been humiliating. He was looking forward to the swing through his home state of Texas to show the president he needed him. Kennedy hoped that Johnson could help repair the rift in the Democratic Party in Texas between Johnson's old friend Texas governor John Connally and Senator Ralph Yarborough, the head of the Democratic Party's liberal wing. But things started out badly at the very beginning of the two-day Texas trip when Yarborough refused to ride in the same car with Johnson. Johnson and Lady Bird intended to impress the Kennedys with a relaxing couple of days at their ranch in Texas. Lady Bird had gone down a week early to prepare for the Kennedy visit. The Kennedys' every whim was catered to: Ballantine's Scotch was ordered for Jack; champagne, Salem cigarettes, and terry-cloth hand towels for Jackie, who, Lady Bird had heard, hated linen ones. Lady Bird herself made up the double bed with the president's plywood board and horsehair mattress for his bad back. They even had a Tennessee walking horse at the ready in case the Kennedys wanted to go for a ride—they knew how much Jackie sought sanctuary in riding.

But the Kennedys would never make it to the ranch. On November 22, 1963, Vice President Johnson and Lady Bird were riding in the third car in Kennedy's motorcade traveling through downtown Dallas. At 12:30 p.m. the car turned onto Elm Street past the Texas School Book Depository traveling at eleven miles per hour, and Johnson could hear the roar of the crowds cheering for the president. Seconds later he heard a cracking sound that he said "startled" him. It sounded like a firecracker or a motorcycle backfiring. Three seconds later a second loud blast could be heard. The first shot hit the president in his back and exited through his neck, the second tore through his skull. Jackie climbed on the trunk of the limousine reaching for a piece of her husband's skull. She went back into the limo and held Kennedy and cried, "Oh, my God, they have shot my husband. I love you, Jack." A half-block behind, the Johnsons could not see what was happening, but Special Agent Rufus Youngblood shouted, "Get down! Get *down*!"

Another shot rang out (according to the Warren Commission there were three shots fired), but by then Johnson was on the floor of the backseat of the car with Youngblood shielding him. Youngblood yelled to the driver to follow Kennedy's car as it raced through the streets to Parkland Memorial Hospital. "Stay with them—keep close!" Youngblood shouted to the driver. When they got to the hospital a few minutes later, Youngblood issued directions: "I want you and Mrs. Johnson to stick with me and the other agents as close as you can. We are going into the hospital and we aren't gonna stop for anything or anybody. Do you understand?" Johnson, who was so often emotional and unpredictable under normal circumstances, replied uncharacteristically calm, "Okay, pardner, I understand."

At Parkland the Johnsons were rushed into a small room where they waited as the president and Texas governor John Connally, who was also shot, were being treated in separate trauma rooms. "Lyndon and I didn't speak," Lady Bird said. "We just *looked* at each other, exchanging messages with our eyes. We knew what it might mean." Confusion took hold and conspiracy theories began to take shape as people tried to find out who had shot the president. The Secret Service thought it the wisest move to get Johnson out of Dallas as soon as possible. Johnson refused to leave until he could find out what happened to Kennedy. Twelve minutes after being wheeled into Trauma Room 1 Kennedy was pronounced dead. At 1:20 p.m. the president's close friend and adviser Ken O'Donnell, a founding member of Kennedy's close-knit circle of friends, the so-called Irish mafia, walked into the small curtained room where the Johnsons were waiting. "He's gone," O'Donnell told them. Johnson would become the thirty-sixth president of the United States and the first Southern president since Andrew Johnson succeeded Abraham Lincoln after his assassination in 1865.

Johnson and Lady Bird were rushed to Air Force One, which was idling on the tarmac at Love Field. Once on board—the plane's

Secret Service code-name was *Angel*—Johnson immediately took charge, though he had never been on Air Force One in flight because Kennedy had always refused. Johnson asked his secretary, Marie Fehmer, for a cup of hot vegetable soup. "I've lived a week since this morning," he told her. Johnson aide Jack Valenti was struck when his boss requested a glass of ice water with "more ice than water" and when it was brought to him he reached for it with a calm, steady hand. The cabin of the Boeing 707 was stifling because the mobile air-conditioning unit was disconnected and the pilot had decided that only one engine should be on in order to conserve fuel. The power was mostly turned off, few lights were on, and the shades were drawn out of fear that the president's death was part of a wider conspiracy to kill everyone at the top of the government. The Johnsons spent a couple of minutes in the bedroom, the same room where Jackie had gotten ready before deplaning, with two single beds and a painting of a farmhouse on the wall. They left to sit in the stateroom instead; it was all too surreal. When Johnson entered the main cabin of the plane three Texas congressmen rose and Congressman Albert Thomas told him, "We are ready to carry out any orders that you have, Mr. President." Lady Bird had not felt the weight of what was happening until, on the drive to Love Field, she noticed a flag at half-staff. As Johnson considered his first moves as president there was a hammering sound as two rows of seats were being removed in the rear of the plane to make room for the former president's casket.

Johnson called Kennedy's mother, Rose, from the plane's conference room. "Mrs. Kennedy," he said softly, "I wish to God there was something I could do for you." He passed the phone to his wife. He then got to the business of governing and called national security adviser McGeorge "Mac" Bundy and his closest aide, Walter Jenkins, and told them to schedule a Cabinet meeting for the next day, Saturday. "Get hold of the congressional leaders. Tell them I want to meet with them as soon as I get to the White House." Johnson was being told by the Defense Department to go back to Washington as soon as possible and fighter planes were being scrambled to escort Air Force One home. But he would not give the order to go knowing how bad it would look to leave without Kennedy's body.

Sid Davis, a correspondent with Westinghouse Broadcasting, was one of three reporters in the motorcade and would recall the scene at Love Field. Davis could not forget the look on the faces of shattered Kennedy staffers. "I saw a lot of the press office staff. These were young people who made the long march with him and won the victory and now all was lost. They were in tears."

Johnson placed another difficult call to Kennedy's brother, Bobby, the attorney general, who was at his Hickory Hill estate outside of Washington. Johnson expressed his sympathy and asked Bobby if anyone had taken responsibility. Then he asked Bobby, who was reeling from news of his brother's murder, if he had the wording for the oath of office. It was a shockingly insensitive request. Bobby said he would have to call him back. Johnson was already president but he felt that taking the oath was a necessary formality to calm the country and provide a sense of continuity. Johnson called Deputy Attorney General Nicholas Katzenbach and asked him to read the oath as it is written in the Constitution as Kennedy listened in and his secretary typed it up.

Meanwhile, Jackie rode with her husband's body in the back of a hearse, and after boarding the plane she went straight to the bedroom. There she found Johnson, his secretary, and his Secret Service agent Rufus Youngblood. They left and headed to the stateroom, embarrassed that the woman who was first lady just hours before had discovered them in the room that was still hers to use. For a brief time Johnson and Lady Bird returned with Jackie to the bedroom and sat next to her on a bed. No one knew what to say. In her diary Lady Bird revisited that moment. "Oh, Mrs. Kennedy," she told her. "You know we never even wanted to be vice president and now, dear God, it's come to this."

Jackie said, "Oh, what if I had not been there. I was so glad I was there."

"I don't know what to say," Lady Bird said. "What wounds me most of all is that this should happen in my beloved state of Texas."

Lady Bird gently asked if she could get someone to help Jackie change out of her bloodstained pink suit. "No," Jackie said. "Perhaps later I'll ask Mary Gallagher [Jackie's personal secretary], but not right now. I want them to see what they have done to Jack." Lady Bird later wrote how alarming it was to see "that immaculate woman, exquisitely dressed, and caked in blood." When she was alone, Jackie lit a cigarette.

Johnson instructed Kennedy aide Ken O'Donnell to ask Jackie to come to the cabin for the swearing-in. "You can't do that!" O'Donnell told him. "The poor little kid has had enough for one day, to sit here and hear that oath that she heard a few years ago! You just can't do that, Mr. President!"

"Well, she said she wanted to do it," Johnson replied. And it was true, Jackie knew she would have to bear witness.

Johnson made it clear he would not accept no for an answer. "We're going to have the swearing-in and I want all you people over here because I want to leave room for the Kennedy people because I want them to observe this." Twenty-seven people jammed into the small and stifling stateroom to witness the ceremony. They needed a Bible but could only find a Catholic missal, or prayer book, in the nightstand in the bedroom. "There's no question in [my] mind," O'Donnell said later, that Johnson "was afraid somebody was going to take the thing away from him if he didn't get it quick." Johnson recognized how important it was to take the oath of office to stem the rising fear. He got his friend Judge Sarah T. Hughes to swear him in and he made sure White House photographer and former army captain Cecil Stoughton was there to record it. Stoughton stood on a stool against the bulkhead to capture the iconic moment.

The ceremony lasted less than thirty seconds. Lady Bird, who was suddenly first lady, thought to herself: *This is a moment which is altogether dreamlike, because the thing is so unreal; we're just like charac-*

At President Eisenhower's second inauguration, Ike's grandson, David, can't take his eyes off Vice President Nixon's daughter Julie, whom he would later marry.

"Ike had the military attitude toward a subordinate," said Earl Warren, whom President Eisenhower appointed chief justice to the Supreme Court in 1953. "And Nixon was a subordinate." Here Eisenhower welcomes Nixon and his wife, Pat, back from one of their most harrowing trips abroad in 1958.

Being a stand-in was agonizing for Lyndon Baines Johnson, a man who had once ruled the Senate. "Every time I came into John Kennedy's presence," Johnson said, "I felt like a goddamn raven hovering over his shoulder."

In a desperate and misguided attempt to win over the Kennedys, Johnson presented their young daughter, Caroline, with a horse named "Tex."

"Shit, shit, shit," muttered John Kennedy's brother Bobby after Kennedy told him he was picking Lyndon Johnson, whom he detested, as his running mate. Here the two sworn enemies attend the groundbreaking ceremony for the John F. Kennedy Center for the Performing Arts, a year after Kennedy's assassination.

Johnson insisted that his vice president, former Minnesota senator Hubert Humphrey, stay out of the headlines, even barring members of the national press from accompanying him on out-of-town trips.

Richard Nixon's vice president Spiro Agnew is the only vice president to resign in disgrace. Less than a year later, Nixon became the first and only president to ever resign. Dick Cheney, who worked for Nixon, described the relationship between the two men as a "train wreck."

From left to right: Secretary of State Henry Kissinger, President Richard Nixon, Vice President–Designate Gerald Ford, and Chief of Staff Alexander Haig meet in the Oval Office to discuss Ford's nomination to replace Agnew.

The day of the primetime address announcing his resignation, Nixon spent twenty-minutes talking privately with his then–vice president, Gerald Ford. The two men had known each other since 1949. As Ford rose to go, Nixon said, "This is the last time I'll call you Jerry, Mr. President."

Ford's vice president, Nelson Rockefeller, was an heir to the Rockefeller fortune, and he was the first vice president who was eligible to live in the Observatory, but he used the house mainly to entertain and preferred to stay at his luxurious Washington estate instead. Gerald Ford and Happy Rockefeller (*left*) with Betty Ford and Nelson Rockefeller (*right*) at the Naval Observatory.

Donald Rumsfeld (*left*) and Dick Cheney (*right*), then young aides to Gerald Ford, were instrumental in kicking Rockefeller off the ticket in 1976. "He had a tendency to be a little bullyish," Rumsfeld said. "He was used to having people agree with him, or he hired them, or he rolled over them."

On November 2, 1976, dark horse candidate Jimmy Carter and his running mate, Walter Mondale, narrowly defeated Gerald Ford and his running mate, Bob Dole. Ford had picked Dole to replace Rockefeller and it remained one of his biggest regrets. "It was one of the few cowardly things I did in my life," he declared.

Walter Mondale (*second from right*) and Jimmy Carter (*right*) had such a close working relationship that most of Mondale's successors, Democrats and Republicans, have sought his advice.

The Naval Observatory has been the official vice presidential residence since 1974.

While Reagan and Bush got along, their spouses did not. Nancy Reagan and Barbara Bush had such a cold relationship that Barbara barely set foot in the White House family quarters over those eight years that her husband was vice president. Here the couples are at the Naval Observatory in 1981.

Bush's top aides did not know he was picking Dan Quayle as his running mate until the day of the announcement. Bush did not even tell Reagan until that morning.

The Bushes at the Naval Observatory the day after Bush won the 1988 presidential election, making him the first sitting vice president to do so since Martin Van Buren.

President-Elect George Bush and Barbara Bush and Vice President-Elect Dan Quayle and Marilyn Quayle visit the Reagans at the White House the day after the election.

ters in a play; this is the beginning of something for us that's dreadful and heavy, and you don't know what it holds. Jackie wanted to be with her husband and spent the rest of the flight with the casket.

When Air Force One landed at Andrews Air Force Base, Bobby Kennedy was waiting out of sight, hunched over in an army truck, avoiding the swarm of reporters gathered for the plane's arrival. He quickly got on board the plane and rushed down the cabin aisle, looking through everyone who stood in his way, including the new president, to console Jackie. He did not say a word to Johnson. A few minutes later, at 6:10 p.m., Johnson spoke his first public words as president: "This is a sad time for all people. We have suffered a loss that cannot be weighed. For me, it is a deep personal tragedy. I know that the world shares the sorrow that Mrs. Kennedy and her family bear. I will do my best. That is all I can do. I ask for your help—and God's."

When Johnson got to the White House, according to his aide Jack Valenti, he did not go to the Oval Office; instead he entered the Diplomatic Reception Room and walked through the basement to the West Wing portico. From there he walked across West Executive Avenue, a guarded street between the White House and what is now called the Eisenhower Executive Office Building, and took the elevator to his second-floor vice presidential offices. There he found Bundy waiting. Johnson met with other important advisers, including Averell Harriman, who was undersecretary of state for political affairs. Afterward, Johnson spoke by phone with President Truman, President Eisenhower, and Sargent Shriver, Kennedy's brother-in-law.

The Johnsons' younger daughter, Luci, was sixteen years old and sitting in Spanish class at Washington's National Cathedral School when she learned about what happened in Dallas. She knew the president had been killed but nothing else. "No one ever said a word about my father or mother," she said. She and her classmates were dismissed, and as she walked into the school's courtyard she saw a member of her father's Secret Service detail. She ran in the other direction, afraid of what he might tell her. He reached for her hand and said, "I'm sorry, I'm so sorry, Luci." He never said that the president was dead, she recalled, "because the words were just unsayable." Once she learned that her parents were all right her mind raced to the next step: their lives would never be the same again.

"President Kennedy was our president but he was also my father's boss," she said, describing a different relationship between her family and the Kennedys. "He also had befriended me and so had Mrs. Kennedy. She had invited me to the White House and made me feel valued and important. Every adolescent is trying to figure out how they fit into the greater scheme of things. She had done a lot to make me feel like I had a place." Luci bought a toy fire engine for John Kennedy Jr.'s third birthday, which was the day of his father's funeral. "She [Jackie] saw my attempt as what it was, just an attempt of a young girl trying to show some love and empathy."

Lyndon Johnson wrote letters to John Kennedy Jr. and Caroline that night. His letter to John, written at 7:20 p.m., one hour and fifteen minutes after Air Force One touched down in Washington, shows how complicated his ascendancy to the presidency was. Johnson had wanted to be president ever since he was a little boy, but not this way. "Dear John," he wrote. "It will be many years before you understand fully what a great man your father was. His loss is a deep personal tragedy for all of us, but I wanted you particularly to know that I share your grief—You can always be proud of him—Affectionately, LYNDON B. JOHNSON."

Shortly after 9:00 p.m. he went home to The Elms, his private residence in northwest Washington, with aides Bill Moyers, Cliff Carter, and Valenti. Lady Bird went upstairs with her press secretary Liz Carpenter, called her daughter Lynda Bird, and changed clothes. "How do you feel?" Lady Bird asked Carpenter. "Chilly," Carpenter replied. "I'm *freezing*," Lady Bird said as she heaped blankets on herself. The day was too much to bear. She sat on the bed to watch one of several television sets scattered around their otherwise elegant home. Bess Abell, the Johnsons' social secretary, re-called going to visit Lady Bird the morning after the assassination. "There was this wonderful little French room off the foyer, it was one of those rooms that Mrs. Johnson liked because there was only one door and she craved privacy. I remember her putting her arms around me and saying, 'Oh Bess, what you've been through.'" It seemed absurd, considering what *she* had been through. But Lady Bird was moving forward.

There had been no time to install secure phone lines before they returned home. As vice president Johnson had no need for them because he had so little responsibility. Two days after the assassination, on November 24, 1963, Johnson asked his friend and speechwriter Horace Busby, or "Buzz," to come to The Elms. That night Busby dutifully sat beside Johnson's bed as he struggled to sleep. At about 10:00 p.m. Lady Bird got into bed. It seemed like a good time for Busby to leave, but Johnson said, "Now, Buzz, don't you leave me. I want you to stay right there till I go to sleep." The room got quiet until Lady Bird sat up and exclaimed, "Lyndon, I just can't stand it." There was less than a year left before the 1964 election, and Johnson and Busby were plotting his 1964 campaign and beyond, and now that they had the presidency Lady Bird was not so sure she wanted it. "Bird, you're just going to have to stand it." Several times, just when it seemed like Johnson had dozed off he would wake up and ask, without opening his eyes, as though just checking his commitment to the job, "Buzz, Buzz, are you still here?" He did that until Busby finally left at 2:00 a.m.

After the funeral, Johnson aides said, Jackie and her children were given plenty of time to move out of the White House. Her one request was to allow Caroline to continue attending the kin-dergarten she had set up in the third-floor Solarium. But the story is different from Kennedy's staff: according to Gustavo Paredes, the son of Jackie's maid, Providencia, "Everybody was scrambling to get stuff out of there . . . they wanted her out, out, out. My mother was scrambling packing up the stuff. OUT." The transition for the

Johnsons was nothing like the Kennedys'—there was no jubilation, no inaugural festivities. At Jackie's request the house was draped in black cambric so that it would look like it had nearly a hundred years earlier after Abraham Lincoln's assassination. The Johnsons moved into the White House on December 7 and the trauma of the assassination was everywhere they looked. "The main thing I remember was how black it all was," Lady Bird recalled. "The White House was full of beautiful chandeliers, but they were all swabbed in black net. Everywhere I looked, the house was draped in black." Lyndon Johnson ordered the fabric taken down on December 22, one month after the assassination.

Johnson asked Kennedy's staff to stay on for a year and most of them, even Ken O'Donnell, agreed. Lady Bird Johnson understood how hard it would be to gain the loyalty and respect of the West Wing aides who had loved Kennedy. In an effort to win over Kennedy aides who were not fans of Johnson's, Lady Bird and the Johnsons' cook, Zephyr Wright, walked through the halls of the West Wing with a grocery cart, passing out warm homemade bread wrapped in aluminum foil and tied with a neat ribbon.

———

When Gerald Ford first heard about Watergate, he was campaigning in Michigan for reelection to his House seat. In the early-morning hours of Saturday, June 17, 1972, several burglars with ties to President Nixon's reelection campaign were caught breaking into the Democratic National Committee offices at the luxury Watergate hotel and apartment complex in Washington, D.C., where they tried to steal documents and wiretap phones. When Ford caught a glimpse of the story in the papers that following Monday—it was not yet the massive scandal that would lead to the first presidential resignation in American history—his first reaction was, "Who would be so stupid as to do something like that?"

But Ford would soon find himself in the middle of the story when, in 1973, Nixon picked him to replace his first vice president,

Spiro Agnew, who resigned from office because of tax evasion and corruption. Nixon assured Ford time and time again that he had nothing to do with Watergate and insisted that he was far too busy to be involved with his reelection campaign (Committee to Re-Elect the President, ironically and mockingly referred to as CREEP): "I didn't pay any attention to the details of my campaign. I was involved with opening the door to the People's Republic of China. I was involved in negotiations with [Soviet leader Leonid] Brezhnev in Salt II. I was trying to end the war in Vietnam. I had turned all of this over to other people."

Ford was suspicious but later said he did not want to believe the president knew about the break-in because he had no desire to take on the presidency if there was an impeachment trial. One afternoon in 1974, Senate Majority Leader Mike Mansfield and Senate Minority Leader Hugh Scott asked Ford to meet them in a smoke-filled office on Capitol Hill. The meeting was confidential and its purpose was to discuss exactly what would happen if Nixon was impeached. "We never asked for an opinion from Ford," Mansfield said, "but we told him, as the leaders of the Senate, that it would be our intention, providing the House impeached him [Nixon], to follow the rules, and we were most seriously considering opening it up to television because it was of such national importance and such international significance."

On the afternoon of August 1, 1974, White House chief of staff Al Haig was upset when he went to the vice president's office. Haig sat on a Davenport sofa with Ford beside him. Haig told Ford that there was information that if verified would be delivered to straight-talking federal judge John Sirica on Monday, August 5, and it would be "catastrophic" to Nixon. The newly discovered tapes of private conversations recorded by Nixon in the Oval Office would become known as the "smoking gun" because they proved that Nixon had tried to obstruct justice by covering up the break-in. The June 23, 1972, Oval Office conversation between Nixon and his chief of staff H. R. "Bob" Haldeman revealed that

Nixon was part of the cover-up. "When that conversation [with Haig] ended, it was obvious to me that either he would be impeached, and I would become president; or he would resign," Ford recalled. "And for the next four or five days the pendulum would swing back and forth, and nobody knew for sure what Nixon's decision would be. I think Betty and I in our own minds knew it was going to take place, and we reassured one another that although we hadn't planned it, and hadn't expected it, we were prepared to do our best." He had been vice president for only eight months.

The next morning, August 2, Ford met with Nixon's lawyer, James St. Clair, who told Ford, "I just want to alert you that things are deteriorating, the whole ball game may be over, and you better start thinking about a change in your plans, a change in your life. I just can't tell you what's going to happen. But I've got to tell you what I know." Ford did not want to weigh in too much with his opinion of what Nixon should do because, if he advised him to resign, it would look like he was shamelessly angling for the presidency. He got home around 7:00 p.m. that night and he and Betty had a dinner to go to. "Things have really totally changed. I want to talk to you," he told her. "We certainly can't talk about it until we get home tonight." He went for a swim, got changed, and then went for dinner at the house of *Washington Star* society columnist Betty Beale and her husband, businessman George Graeber. The Fords got home around 11:00 p.m., made their usual bourbon and water nightcap, and sat in the family room. Ford told Betty that either he would become president in a matter of days or "we would be dangling for six months while an impeachment process went through in the House and an impeachment trial went through in the Senate."

They spoke for more than an hour that night, and it was a painful conversation. They were friends with the Nixons, and Betty worried about how Pat was handling her husband's public downfall. Betty was not outwardly angry at Nixon, Ford said. "I think she was more overwhelmed with sadness on the one hand vis-à-vis the Nixons, and wondering how things would develop as far as we were concerned." Betty told her husband, "My God, this is going to change our whole life." Ford got a call late that night from Haig, who said the situation had not changed. At midnight the Fords prayed in their small bedroom together. They had twin beds with a single headboard, and with Ford on the right and Betty on the left he reached out his left hand and clasped hers. Every night Ford said a particular prayer that he used to recite when he was at sea in World War II. On this night though they prayed for guidance.

"I was real scared," Betty recalled. "I think my thoughts were more toward the responsibility as a woman and a wife and a mother, 'How was it going to affect the family?' As wives and mothers we tend to project as 'what's going to happen to everybody else?' And not 'how is it going to affect us?' That's one of our failings I think. We forget to consider our own impact."

Just the week before Nixon resigned, the Fords were getting ready to move into the official vice presidential residence at One Observatory Circle, about three miles from the White House. When he and Betty arrived at the Victorian-style house to talk with interior decorators about the décor, Ford already knew that the president would be resigning in a matter of days. Betty had already selected the china, cobalt blue and gold trim with the vice presidential seal along the border. When they finally had a moment alone he whispered to her, "Betty, we are never going to live in this house."

Before 8:00 on the morning of Wednesday, August 7, 1974, Haig met again with Ford in his vice presidential office in the Executive Office Building. He came with a clear message after all the months of questions: "Mr. Vice President, it is time for you to prepare to assume the office of President." Ford summoned his closest advisers—Robert Hartmann, Jack Marsh, and Philip Buchen—to begin the transition. At 11:15 Ford went to Capitol Hill for his usual Wednesday prayer meeting with old friends from Congress and former defense secretary Melvin Laird. Ford could not tell them what he knew, but they offered prayers for Nixon's successor—everyone in the room knew that it would be Ford. Ford later summed up the situation: "It struck me that the ball game was in the bottom of the 9th, and we didn't have anybody on base, and we had two outs, and we didn't have much chance of prevailing."

The next day, August 8, Ford was summoned to the Oval Office at 11:00 a.m. When he walked in alone he found Nixon and Haig. Nixon told Haig to get White House photographer Ollie Atkins and Ford knew then that Nixon wanted a picture to record the historic meeting. Nixon asked Ford to sit down and told him, calmly, "Jerry, I am resigning. You will become President. I know you'll do a good job. I have complete faith that you will carry out my foreign policy, that your views and mine will continue to be similar on domestic policy." During their twenty-minute conversation, Nixon urged Ford to keep Henry Kissinger on as secretary of state. "Henry is not the easiest person to work with, but he is an outstanding foreign policy strategist . . . and you and the country need him." Nixon's face was drawn after only three hours of sleep, and the two men reminisced about their families' long friendship. They had known each other since 1949, when Ford was first elected to Congress and Nixon had been first elected two years earlier. That night, in a televised prime-time address, Nixon would announce his resignation.

As Ford rose to go, Nixon said, "This is the last time I'll call you Jerry, Mr. President." They shook hands and Ford left with the feeling that Nixon felt victimized by the circumstances. Nixon told Ford the transition would happen the next day when he was flying over St. Louis. Years later Ford said the whole situation just "tore at your insides," but that it was complicated because he also felt betrayed by Nixon for lying to him about Watergate. Betty described August 9, 1974, the day the Nixons left Washington and she became first lady, as "the saddest day of my life."

Connie Stuart was Pat Nixon's chief of staff and press secretary at the time. She remembers thinking Pat seemed relieved when her husband resigned after living under a cloud of deepening suspicion for two years. "They were so shell-shocked and ripped apart," Stuart remembered. Up until her death in 1993, Pat maintained that no one knew the full story of Watergate. She knew intellectually that her husband had done something wrong, but she did not want to acknowledge it to herself. Stuart compares Pat's willful ignorance to that of a woman whose husband has been having a long affair. Pat was such a loyal defender of her husband that she cut out news articles about other presidents, such as President Franklin Roosevelt, who were said to have bugged the White House, and set them aside. Nixon had done nothing uniquely wrong, she reasoned. All that rationalizing was exhausting, Stuart said. The end was almost a relief. "It's kind of like having a sick parent who's dying, you want them to live, because they're your parent, but on the other hand it would be so much better if they just went."

On September 8, 1974, a month after Nixon announced his resignation, Ford granted the disgraced former president a "full, free, and absolute" pardon. The decision contributed to his 1976 loss, with one aide likening it to hara-kiri, the Japanese ritual suicide. The Fords' youngest son, Steve, said he remembers warning his father, "They're going to crush you if you pardon Nixon." He said his father used an analogy to explain his decision: "The president is like the father of a family," Ford told his son, "and sometimes the father of the family has to hand out discipline or justice that may seem unfair but it's really done in a way to keep the whole family together."

———————

Section 3 of the Constitution's Twenty-fifth Amendment enables the president to declare himself temporarily disabled, and it has been invoked only three times, allowing the vice president to serve as president: in 1985, when President Reagan underwent surgery to remove cancerous polyps in his colon; in 2002, when President George W. Bush had a colonoscopy; and in 2007, when Bush had a second colonoscopy. In each case the president sent a letter to the president pro tempore of the Senate and the Speaker of the House stating that he was "unable to discharge the powers and duties of the office." The vice president became president until the president sent a letter to those same congressional leaders stating that he could resume the presidency.

But in the most dramatic case when a vice president actually had to assume presidential duties, there were no such formalities. Sixty-nine days into Ronald Reagan's presidency, at 2:25 p.m. on March 30, 1981, a nondescript twenty-five-year-old named John Hinckley Jr., bizarrely obsessed with the movie *Taxi Driver* and the actress in it, Jodie Foster, fired six bullets at President Reagan and his entourage as they left the Washington Hilton hotel. It was part of a misguided attempt to impress Foster. One bullet struck the president in the lung and another struck his press secretary, James Brady, in the head. He also wounded a police officer and a Secret Service agent.

At the time Vice President George H. W. Bush had just delivered a speech at the Southwestern Cattlemen's Association in Fort Worth, Texas, and was on Air Force Two. As the plane was about to take off to fly him to another speech, this one before the state legislature in Austin, his Secret Service agent, Ed Pollard, entered his private compartment. "There was a shooting incident at the Washington Hilton involving the President," he said, "and two agents are down." The president, Pollard told Bush, was not hurt. It was decided that Bush would give the address in Austin and cancel the rest of the day's events and fly back to Washington. But then a second report came through and Pollard told Bush, who remained composed in a navy blue flight jacket and a loosened tie, that the president had in fact been shot. "We'll go right back to Washington," Bush said.

As the rest of the world watched video of the shooting on television, chaos consumed the White House. Emotions ran high. Learning that the gun was a .38, national security adviser Richard

Allen exclaimed, "Jesus Christ!" With Bush on board Air Force Two on a four-hour flight back from Texas, a power struggle emerged in the Situation Room, a long paneled windowless room in the White House basement that is secured behind electronic locks and guarded by uniformed Secret Service agents. Secretary of State Al Haig and Secretary of Defense Caspar Weinberger were there locked in a battle over who was in charge. Simultaneously, aides were being told that Brady was being administered last rites and an incorrect news report pronounced him dead. Reagan aides tried to locate Brady's daughter and bring her to Washington. In the meantime, the Pentagon was reporting more Soviet submarines than usual off the East Coast, and transcripts of the conversation between Weinberger and Haig show a bitter tug-of-war over what to do about it. Weinberger did not want to raise the military's alert condition—DEFCON—and the two men wondered about the location of the "football," a briefcase containing the nuclear release-codes that is always at the president's side. Aboard Air Force Two, Bush told two Texas congressmen who were traveling with him, Democratic Majority Leader Jim Wright and Republican James Collins, "I was thinking, how could anyone have taken a shot at Ronald Reagan?" Bush was confident in his abilities to take over the presidency but was very worried about his friend. He told his executive assistant, "My innermost thoughts are [that] this guy is a friend . . . I think about Nancy Reagan: Is anybody holding her hand?"

Haig gave a bizarre press conference meant to reassure the public, but it only caused more confusion and concern. When asked by a reporter who was in charge, Haig ignored the rules of succession (under the Twenty-fifth Amendment, ratified in 1967, after the vice president the Speaker of the House and the president pro tempore of the Senate are next in the line to succeed the president, ahead of the secretary of state) and said without seeming at all calm, "As of now, I am in control here, in the White House, pending return of the Vice President and in close touch with him."

Bush was on his way back, but he had no means of secure voice communications from Air Force Two. Around 3:30 p.m. White House counsel Fred Fielding asked an aide to draft a letter for the transfer of authority from the president to the vice president pursuant to the Twenty-fifth Amendment. In the flurry of activity the letter was mistakenly addressed to Senate Majority Leader Howard Baker instead of Strom Thurmond, who was the president pro tempore of the Senate. A letter withdrawing that transfer, to be used once the president was no longer incapacitated, was also drafted. But the first letter was never sent and authority was never formally transferred to Bush. Secretaries and attorneys were told not to tell anyone about the drafting of the letters.

Watching the drama unfold aboard Air Force Two, Bush remained calm. Shortly before landing he got a call from Reagan aide Ed Meese, who was at the hospital. After a moment Bush said, "Well, that's wonderful. That's great news." "The bullet has been removed," Bush told everyone on board, "the operation has been a success, and the president is fine." Bush felt strongly about the line of succession and did not want to appear to be taking over prematurely from the president or to cause alarm. The seventy-year-old Reagan, we now know, was much closer to death than the public knew at the time, having lost 40 to 50 percent of his blood volume.

Bush landed at Andrews Air Force Base at 6:30 p.m. The plane taxied into a hangar as an extra level of security. There was some debate about whether Bush should take the waiting helicopter straight to the White House, but he put an end to it. Only the president lands at the White House, he said. The helicopter landed at the Naval Observatory and Bush was driven in an armored limousine to the White House. He gathered Reagan's chief of staff James Baker, Haig, Fielding, national security adviser Richard Allen, treasury secretary Donald T. Regan, domestic adviser Martin Anderson, and David Gergen, who was then an assistant to the president, to accompany him to the Situation Room. Bush and Baker asked if everyone present had the proper security clearance to be there and some people shuffled out of the crowded room. "The President is still the President," Bush told them. "He is not incapacitated and I am not going to be a substitute President."

At 8:30 p.m. Bush said in a televised statement, "I can assure this nation and a watching world that this government is functioning fully and effectively." He went to see Nancy in the private residence (she had wanted to stay with her husband in the hospital but was advised against it because it would send the message that the situation was dire). Bush later told friends that she looked "tiny and afraid." When asked for his favorite memory of his eight years with Reagan, Bush remembered visiting him in the hospital after he had been shot. "When I first entered the room, he was nowhere to be found. He called from around the corner, where I found him on his knees wiping the floor. He had spilled a glass of water, and didn't want the nurses to get into trouble. So thoughtful, so typically thoughtful."

During the following days Bush worked out of his West Wing office and his office in the Eisenhower Executive Office Building. James Baker ordered a review of the administration's response and the Secret Service reviewed and formalized unwritten procedures to increase protection for the vice president and other successors to the presidency in the event of a future attempt on the life of a president.

With Reagan recuperating in the hospital, Bush held a Cabinet meeting and still sat in his usual chair, across from an empty seat at the center of the table—that chair, Bush knew, was always reserved for the president, whether he was there or not.

VII

From Senator to Subordinate: The Story of Nixon/Eisenhower, Johnson/ Kennedy, and Humphrey/Johnson

Ike [Dwight D. Eisenhower] had the military attitude toward a subordinate. And Nixon was a subordinate. He admired George Humphrey, the secretary of the treasury, because he had money, but Nixon had no money and was inferior in rank.

—SUPREME COURT CHIEF JUSTICE EARL WARREN
TO JOURNALIST DREW PEARSON

Humphrey knew LBJ before he took the vice presidency, and he knew he could be a son of a bitch. And he was.

—A FRIEND OF HUBERT HUMPHREY'S

Dwight David Eisenhower, a five-star general and supreme commander of Allied Forces in Europe during World War II, was an almost mythical figure. When he turned seventy in 1960, toward the end of his second term, General Ike became the oldest president to serve in office. (He no longer holds the record as the oldest elected president—Ronald Reagan took the oath of office when he was sixty-nine and Donald Trump was sworn in at seventy.) Eisenhower suffered a series of health emergencies while president, including a massive heart attack in 1955, intestinal surgery in 1956, and a stroke in 1957. His vice president, Richard Nixon, was watching and waiting. After Eisenhower's stroke, Nixon got a call from Eisenhower's top White House aide, Sherman Adams, who said, "This is a terribly, terribly difficult thing to handle. You may be president in the next twenty-four hours."

Well before the Twenty-fifth Amendment was ratified, Eisenhower sent Nixon a memo laying out how Nixon should deal with a situation in which he was unable to serve but was still alive: "The existence of this agreement recognizing your clear and exclusive responsibility for deciding upon the inability of the president to perform his duties and exercise his powers will remove any necessity or desire on the part of friends and staffs to impede the right and authority of the vice president in reaching his decision on the matter." After visiting Eisenhower, who was recovering from one of his many illnesses, Nixon jotted down notes to himself on White House stationery. Eisenhower, he wrote, "looked like [a] corpse . . . waxen face." But Eisenhower insisted, "Drs. say I'm getting better!" Nixon, like most vice presidents, was following the president's health more closely than anyone else.

Richard Milhous Nixon is best known as the first and only president to resign the office, but before then he had a swift rise to power. In six years, he went from a member of the House of Representatives to vice president. He was born in 1913 in Yorba Linda, a small California town, and grew up in East Whittier, another small town in California. He was the second of five brothers and his parents ran a combination grocery store and gas station and struggled financially. Nixon excelled at Duke University's law school and enlisted in the navy during World War II. When he returned after the war he was drafted to run for Congress by a group of prominent southern California Republicans. In 1946 he defeated the five-term incumbent, Democrat Jerry Voorhis, by more than fifteen thousand votes, winning with 56 percent of the vote. GOP leadership rewarded him with a seat on the House Un-American

Activities Committee, where he became famous during the investigation of Alger Hiss, a State Department employee accused of passing secrets to the Soviets. Nixon won reelection in 1948, and in 1950 he won a seat in the Senate. In 1952 Eisenhower, sixty-two, picked Nixon, thirty-nine, as his running mate. Eisenhower, who was used to commanding total respect and obedience, was old enough to be Nixon's father and he often treated Nixon like a junior officer. "There were times," said Nixon's friend, journalist Ralph de Toledano, "when I would find Nixon literally close to tears after a session at the White House during which Eisenhower humiliated Nixon."

Controversy struck during the campaign when Nixon was accused of misusing campaign funds amounting to about $18,000 to offset day-to-day office expenses and the *New York Post*'s headline screamed, "SECRET RICH MEN'S TRUST FUND KEEPS NIXON IN STYLE FAR BEYOND HIS SALARY." Eisenhower weighed the possibility of dropping him from the ticket and the two did not speak for days. Right after the story broke, Eisenhower told an adviser, "I do not see how we can win unless Nixon is persuaded to withdraw." The silent treatment was agonizing for Nixon and he was forced to interpret Eisenhower's public statements as a private rebuke. At a speech to the National Federation of Republican Women's Clubs in St. Louis, Eisenhower repeated phrases like "Honesty is the best policy" and "He that goes a-borrowing, goes a-sorrowing." It seemed like a passive aggressive way to communicate his displeasure with his vice president. When they finally talked, after sixty hours had passed, Eisenhower at first sympathized with Nixon: "You've been taking a lot of heat the last couple of days." But then he directed his running mate to go on national television to explain himself. Nixon agreed but he wanted Eisenhower to promise his support.

"After the television [appearance], General, if you think I should stay on, I think you should say so," Nixon said. "The great trouble here is the indecision."

"We ought to wait three or four days after the television show to see what the effect of the program was," Eisenhower replied, not giving an inch.

A furious Nixon replied with something decidedly less reverential and told the man who led the Allies to victory in World War II, "There comes a time in matters like this, when you've either got to shit or get off the pot."

"Well, Dick," Ike replied, ignoring Nixon's insubordination, "go on the television show, and good luck. Keep your chin up." Nixon's wife, Pat, who never liked politics, was made furious by their exchange. She asked her husband, "Why should we keep taking this?"

Nixon, a good soldier, prepared his so-called Checkers speech, a nationally televised address defending his actions, but a half-hour before he was supposed to go on he received a call from New York governor Thomas Dewey, who was a sometimes intermediary between Nixon and Eisenhower. "I hate to tell you this, Dick," he said. "There has just been a meeting of all of Eisenhower's top advisers and they have asked me to tell you that it is their opinion that at the conclusion of the broadcast tonight you should submit your resignation to Eisenhower."

"It's kind of late for them to pass on this kind of recommendation to me now," Nixon said, before hanging up. The seeds of the paranoia that led to his downfall were sown in those humiliating thirty minutes when he felt truly disposable. But Nixon collected himself, stuck with his original plan not to leave the ticket, and saved his political future in the process. "Pat and I have the satisfaction that every dime that we've got is honestly ours," Nixon declared, referring to his wife. "I should say this, that Pat doesn't have a mink coat. But she does have a respectable Republican cloth coat, and I always tell her she'd look good in anything." The speech got its famous name because he vowed to keep one gift: a black-and-white dog that his daughters named Checkers. It was considered a success, and though at first noncommittal, Eisenhower let Nixon stay on the ticket and they won the election in a landslide.

New York Times reporter James Reston asked Eisenhower, a proud political novice, what his role was in the selection of Nixon and he replied, with surprising honesty, "The first thing I knew about the president or any presidential nominee having any great influence on the vice presidential selection was, I think, about the moment that I was nominated." He said he wrote down the "names of five, or maybe it was six" palatable possibilities. It was important to Eisenhower that his vice president be younger and represent the future of the party. Nixon was on that list but he never was able to garner Eisenhower's respect. Eisenhower campaign aide Herb Brownell called Nixon and said dispassionately, "We picked you." Nixon was summoned to meet Eisenhower in his suite and in a moment of pure exuberance said, "Hi, Chief," when he saw him. At that moment, he later said, he sensed "a little coolness developing" between them. Eisenhower did not like Nixon's giddy tone. At the convention, Nixon grabbed Eisenhower's wrist when they were at the podium, he said years later, "I sensed when I held the general's arm up that he resisted it just a little." Nixon was hopeful that he would play a large role in Eisenhower's White House, and he tried very hard to please the president, taking up golf (Eisenhower's favorite sport) and filling in for the president at social functions. He revered Eisenhower but Eisenhower had a decidedly suspicious view of Nixon, whom he considered a professional politician, which in Ike's eyes did not earn Nixon his immediate respect.

Nixon had high hopes at the beginning of his vice presidency. "He [Eisenhower] has also indicated on several occasions that he believed it is essential that a Vice President know what is going on," Nixon wrote after meeting with Eisenhower in 1953. But Eisenhower demurred and said that though he wanted to give Nixon more responsibility, the law would not allow it. Because the Senate was so closely divided between Republicans and Democrats, he needed Nixon to focus on Congress. "On several occasions," Nixon said, "the President and sometimes a Cabinet officer will ask in a meeting what I believe the congressional viewpoint will be on a certain matter . . . When I became Vice President I thought that one of the greatest services I could render would be to interpret the administration's program to individual members of the House and Senate who might otherwise not receive this information."

Nixon kept in touch with Lyndon Johnson, who became majority leader in 1955 after the Democrats won control of the Senate, and a few key Senate Democrats with whom he had long-standing relationships. Years later Johnson told his vice president, Hubert Humphrey: "Eisenhower cut his guts out and didn't want Nixon to have any power." When Eisenhower was weighing whether to run for a second term he invited friends and associates, including Nixon, to sit with him in the Red Room of the White House. "We must present the able, personable young men of the Republican Party, on television, on radio, on platforms and forums, and do it right away." He continued, with Nixon looking on, "Take Bob Anderson [Eisenhower's forty-four-year-old deputy defense secretary], for instance. That man's got Texas in the palm of his hand. I think he is a perfectly wonderful young man. My God, he could run for Pope on the Presbyterian ticket and get elected. That's the kind of Republican face and voice that ought to be seen and heard." Nixon, who was obviously ambitious and at the time forty-one, could not have ignored the slight.

As Eisenhower's health declined, he relied more and more on his vice president. In a note dated October 1, 1955, as he was recovering from a heart attack in a Denver hospital, Eisenhower wrote: "Dear Dick: I hope you will continue to have meetings of the National Security Council and of the Cabinet over which you will preside in accordance with the procedure which you have followed at my request in the past during my absence from Washington." Well before the Twenty-fifth Amendment was adopted in 1967, Nixon and Eisenhower had a gentleman's agreement that Nixon would take charge while Eisenhower recovered. Even though Nixon did an admirable job, Eisenhower still weighed the possibility of dropping Nixon from the ticket in 1956. He told Nixon that he might get better training for the presidency if he ran the Pentagon instead of being his vice president in his second term. Eisenhower acted as guidance counselor to Nixon and told him, "There has never been a job I have given you that you haven't done to perfection." He added condescendingly, "The thing that concerns me is that the public does not realize adequately the job you've done." As he was considering what to do, Eisenhower excluded Nixon from an important strategy session. "Ordinarily you would be the first one I would ask," Eisenhower explained. But "since you are going to be so much the object of conversation, it would be embarrassing to you." Nixon felt it necessary to outmaneuver his boss and orchestrated a write-in effort in the New Hampshire primary to prove his own popularity. He garnered an unexpectedly high number of votes, which made him more valuable to the reelection.

Still, when Eisenhower announced that he would be seeking a second term, he declined to comment on whether Nixon would be staying on the ticket; he even said that he would have to wait for the Republican National Convention, which was nearly six months away, to decide. Ultimately, Nixon forced himself to remain on the ticket and Eisenhower complied—but not before putting Nixon through the wringer. As he had in 1952, Nixon served as Eisenhower's hatchet man and was dispatched on grueling campaign trips. During the 1954 midterm elections, between mid-September and Election Day, he traveled more than twenty-five thousand miles and visited nearly a hundred cities in thirty states. But Democrats won control of both houses of Congress, despite Nixon's efforts. During the 1956 reelection campaign Nixon was sent on a thirty-two-state tour of the United States. He dutifully praised Eisenhower on the stump, calling him "a man of destiny, both at home and abroad," and "a man who ranks among the greatest of the legendary heroes of this nation."

As vice president, Nixon set aside his own ego and thirst for power. "The first responsibility of a vice president," he said, "is loyalty to the President and I have some very definite ideas, for example, about defense, about organization of government generally, but I prefer not to express them at this point." He desperately wanted to be president, but as vice president he had to check with the president or his staff before doing anything. "Once the decision is made as to what the Administration's program is . . ." he said, "all members of the Administration should either support that decision or get out."

As his health deteriorated, Eisenhower asked the Nixons to take an eighteen-day diplomatic trip to South America in the spring of 1958. The point of the trip was to celebrate the inauguration of Arturo Frondizi, who was the first democratically elected president in Argentina in two decades. The visit was going well until the Nixons arrived at the University of San Marcos in Lima, Peru, where rocks were thrown at them by leftist demonstrators. Things got worse during a later stop in Caracas, Venezuela, when the Nixons arrived at the airport and protestors spat on them and threw fruit and garbage at them. Protestors blocked their route with a vehicle, and their motorcade was under attack as rocks and pipes were thrown at their car. Nixon's wife, Pat, was not sure if they would survive. A rock struck the vice president's window and a piece of glass hit the foreign minister's eye and he started to bleed. The demonstrators began rocking the vice president's car, trying to overturn it. Secret Service agents did not want to draw their guns for fear it would cause more violence. After more than ten minutes, agents were able to use a press car to block traffic and give the Nixons' motorcade a path to speed away and escape to the American embassy.

The next day members of the press gathered around the cars that Nixon insisted be left in full view so that their harrowing journey could be documented. Several American reporters burst into spontaneous applause when the Nixons left the embassy to attend a government luncheon. Tears welled up in the normally stoic Pat's eyes. The Nixons were welcomed home as heroes. The Eisenhowers met them at Andrews Air Force Base, along with thousands of supporters, half of Congress, and the full Cabinet.

Even after all that, Nixon was never brought into Eisenhower's inner circle. He and Pat were never invited to the president's Gettysburg Farm when he was vice president, and they were never even invited to the White House residence for a party. "Ike had the military attitude toward a subordinate," Chief Justice Earl Warren, whom Eisenhower had appointed to the Supreme Court, told journalist Drew Pearson. "And Nixon was a subordinate. He admired George Humphrey, the secretary of the treasury, because he had money, but Nixon had no money and was inferior in rank."

In 1960, Nixon sent a birthday note to Eisenhower in the final days of his campaign for the presidency against John F. Kennedy. "When I talked to you on the telephone it was 6:30 am California time and I failed to realize how significant this particular day was for you and for the nation," he wrote timidly. "Pat joins me in sending our very best to the man who is oldest in years and youngest in spirit ever to grace the White House.—Dick Nixon."

Eisenhower did not endorse his vice president until he was nominated at the Republican convention. When a reporter asked Eisenhower at a press conference if he would endorse Nixon, he replied, "There are a number of Republicans, eminent men, big men that could fulfill the requirements of the position." At an August 24, 1960, press conference, a month after the convention, Eisenhower was asked if he could point to a specific accomplishment of his vice president. He replied warily, "If you give me a week, I might think of one. I don't remember." A week later, he still could not offer a single example.

Eisenhower's shadow never left Nixon. During his 1968 campaign for president (he lost in 1960), Nixon asked his aides to always keep Eisenhower updated on how he was doing in the polls. He wanted to make sure that Eisenhower knew when he was doing well. And he all but begged Eisenhower for his endorsement, which he finally gave less than four months before Nixon won the election. "After Nixon got elected, he invited a few of us to go upstairs and see what the living quarters were like," said Nixon's political strategist John Sears. "And all he talked about was how it looked when Eisenhower was president—where he had his medals and what he had over there and where he had his desk."

———————

John F. Kennedy narrowly defeated Nixon in 1960 in one of the closest presidential races in U.S. history. Kennedy and his vice president, Lyndon Johnson, could not have been more different, but Johnson, who prided himself on his ability to size people up, had a grudging respect for Kennedy. After a few months in office, he told *New York Times* reporter Russell Baker how much he admired Kennedy. The president, Johnson said, "looks you straight in the eye and puts that knife into you without flinching." It would be a backhanded compliment to some, but for Johnson it was genuine praise. In the early days of their partnership they behaved like a married couple. Kennedy's secretary, Evelyn Lincoln, was with Kennedy at the family's Palm Beach compound before the inauguration, and during that period Johnson visited three times. She arrived at the house early one morning from her hotel to find Johnson sitting at one end of the large dining room table in a mono-grammed "LBJ" terry-cloth robe (to Johnson it was a sign that he had risen in stature if he could be known by his initials, like FDR) with the *New York Times*, Kennedy's favorite newspaper, sitting, untouched, by his side.

"Good morning, Lyndon, you're up early this morning," Kennedy, who was also wearing a terry-cloth bathrobe, said when he sat down for breakfast at the opposite end of the table. When he noticed the *New York Times* was missing, he glanced up and saw it sitting on the table next to Johnson, who apparently was not interested in reading it. Kennedy was not amused. Johnson energetically talked to Kennedy throughout breakfast. Later, when Lincoln went to bring Kennedy a cup of coffee in his bedroom, the president-elect said, "What gall. If he had the paper why didn't he read it? Or at least give someone else a chance to see it."

Johnson wielded tremendous power as Senate majority leader and was so desperate to cling to it that, in a remarkable move, he had persuaded the Texas state legislature to pass a law that allowed him to run for reelection to his Senate seat at the same time that he was running for vice president. After he was elected vice president, Johnson tried to keep hold of his power over the Senate by asking Mike Mansfield, the senator who would take his place as majority leader, to allow him to continue as conference chairman so that he would be in charge of Senate Democrats whenever they conferenced. It would have been an unprecedented consolidation of power. The mild-mannered Mansfield was amenable to the arrangement, but Senators Robert Byrd and Richard Russell refused to agree to such a breach of the constitutional separation of powers that would give a member of the executive branch too much influence over the legislative branch.

———————

Johnson's impression that Eisenhower "cut his [Nixon's] guts out" is ironic since Kennedy did far worse to him. He allowed Johnson to be routinely humiliated. Johnson never liked being first in line. Being a stand-in was less than satisfying for a man who had once ruled the Senate. His base of conservative Southern Democrats abandoned him once he joined the administration—they thought Kennedy was too liberal. His less than three years as vice president were agony for him. "Every time I came into John Kennedy's presence, I felt like a goddamn raven hovering over his shoulder," Johnson said after leaving the White House. "Away from the Oval Office, it was even worse. The vice presidency is filled with trips around the world, chauffeurs, men saluting, people clapping, chairmanships of councils, but in the end, it is nothing. I detested every minute of it."

Johnson was born in 1908 in central Texas. He grew up with

little money and became passionate about civil rights in part because of his experience teaching students of Mexican descent in Cotulla, Texas. He ran and won a seat in the House in 1937, and after six terms he was elected to the Senate in 1948. In 1953 he became the youngest minority leader in the Senate, and the following year, when Democrats won control of the Senate, he became majority leader. He used his imposing physical presence—he was six-foot-four—and his larger-than-life personality to convince his colleagues to vote for legislation, often looming over them. Becoming vice president was in many ways a step down for Johnson because he had wielded so much power in the Senate.

Never one to do things halfway, in 1960, Johnson threw himself into the presidential campaign. Tyler Abell, who was on his advance team and charged with making sure that crowds showed up at rallies, recalled how Johnson did "whatever it took" to get Kennedy elected. He never wanted to get off the stage. He would be giving a speech and ignore the alarm on his wristwatch for several minutes until he would finally tell the crowd: "Lady Bird keeps tugging at me and says I should stop, so I guess I better stop." The first advance trip Abell did was in Boston, where "nobody had ever heard of him." Johnson arrived on a weekday afternoon and Abell was surprised to see people come running out on the street to see his motorcade pass by. He soon discovered why. Johnson had a trick—he brought thousands of Senate passes to hand out to the crowd. "He had suitcases full and handed them out to everybody. People came flocking down from the buildings out into the street." Boston was a distance from Washington, but it was a personal, and even a little bit hokey, touch that they appreciated. Lyndon Johnson may not have been a household name, but he intended to use his place on the ticket to become one.

During one motorcade, Johnson got out of his car and approached a mounted police officer and asked him to get off his horse. LBJ then demonstrated to the crowd his experience growing up in Texas, placing his foot in the stirrup and hoisting himself up.

"Sitting in the saddle," Lincoln wrote in her memoir, "he began to talk to the crowd—about our foreign policy!" Kennedy's staffers might have been laughing at him, but there was no denying that Johnson helped Kennedy win the South and that, because Kennedy was Catholic, Johnson's presence on the ticket was important. Even Bobby Kennedy gave Johnson and his wife, Lady Bird, credit for helping his brother win the South.

With his power stripped away as vice president, Johnson obsessed over the smallest things. There was no official vice presidential residence then, so the Johnsons lived at The Elms, their sprawling mansion that once belonged to socialite Perle Mesta in Washington, D.C.'s leafy Spring Valley neighborhood. Johnson called his assistant Bess Abell, Tyler's wife, "the maintenance engineer" of The Elms. Her phone rang so often in the middle of the night with one request or another that she began keeping a small notebook on her bedside table so she could take notes. When Tyler first started working for Johnson, his boss, who had worked for LBJ for years, warned him: "You've got to be very careful. This man wants everything done his way. The soda water has to be Clicquot Club and the scotch has to be Cutty Sark." Once, Tyler recalled, Johnson was sitting by the swimming pool at The Elms where Bess had arranged a swim and a picnic for him and his family. Sitting in his own swimsuit, Johnson got on the phone and called an assistant with an urgent request: he wanted more bathing suits at the house. "I want you to get some bathing suits so people who come to swim at this pool, if they don't have a swimsuit, they'll be able to use one of ours. Get some big ones and get some small ones," Johnson instructed. He had nothing better to do.

The vice presidency was torture for a man who had run the Senate and was used to commanding attention when he entered a room. Kennedy aides turned their backs on him at cocktail parties. "Oh, that's just the vice president," a West Wing staffer said to a party guest who did not even recognize Johnson. Their favorite nicknames for him and his wife, Lady Bird, were "Uncle

Cornpone and his Little Pork Chop." Johnson referred with contempt to the elite Kennedy clan and their aides as "Bostons" and "Harvards." Things got so bad that when Kennedy asked Congress for Secret Service protection for the vice president and his wife, Johnson told his friends, "He [Kennedy] just wants to spy on me." Johnson was well aware of the Kennedy aides' disdain for him and the snickering behind his back.

Johnson sometimes asked to fly with Kennedy. "That's ridiculous," Kennedy told his staff. "He has his own plane. It wouldn't be practical for both of us to be traveling together." Kennedy added, exasperated, "How many times must I tell him that the President and the Vice President, as a matter of security, should never ride on the same plane?" *Whatever happened to Lyndon?* became a running joke among Johnson's former congressional colleagues. Johnson requested an office in the West Wing of the White House, which Kennedy deemed absurd. To make it look like he was included in big decisions, Johnson had his driver drop him off near the sidewalk leading to the Oval Office. He would walk along the colonnade, stroll past the Oval, and enter through a door to the president's secretary's office. "Nearly every morning he would open that door, grunt, and pause for a moment to look around to see what was going on," according to Lincoln. From there he would check to see if Kennedy was in and if he was not he would walk into the hall leading to the reception room and make sure the reporters who were gathered there got a good look at him. When he came out to cross the street to his office in the Executive Office Building, it gave the impression that he had just been in a meeting with the president.

Johnson was further humiliated when, on foreign trips, Kennedy insisted that a State Department official accompany him to make sure that he kept his appointments. The vice president did not trust these officials and thought they were spying on him. When John Glenn became the first American to orbit the earth on February 20, 1962, Johnson wanted credit as chairman of the National Aeronautics and Space Council. He insisted on being included in the ticker tape parade down Broadway with Glenn. Inside the West Wing this was seen as yet another desperate attempt by the vice president to claim credit and further his own career. Johnson aide Bill Moyers felt that Johnson had become "a man without purpose . . . a great horse in a very small corral."

Even after all that, Johnson desperately wanted to be accepted by Kennedy. After a breakfast meeting with members of Congress, Johnson and Kennedy were walking to the Oval Office when Johnson surprised Kennedy with a gift. He would be sending some cattle and a horse named "Tex" for Kennedy's daughter Caroline to their Middleburg, Virginia, estate. Kennedy already had a menagerie of pets, including dogs, birds, a cat, and a rabbit, and he had a rule that he would not accept gifts worth more than fifteen dollars (or the equivalent of $120 today). But he knew he was trapped—rejecting the gift would deeply hurt his already emotionally fragile vice president. When the horse arrived, Kennedy, Jackie, and their young daughter dutifully posed for a photograph on the South Lawn of the White House with a beaming Johnson before "Tex" was sent to the Kennedys' country house. Not long after that, Johnson saw Caroline, who was just four years old at the time, drawing in Evelyn Lincoln's office. "Do you know who I am, Caroline?" he asked. Caroline said nothing. Lincoln leaned down and whispered, "That's Mr. Johnson, Caroline."

"That's right. I'm your uncle Lyndon, remember?" Johnson said. "I'm the one who gave you that fine riding horse, Tex."

Caroline mumbled shyly, "Oh."

"Now remember what I told you, Caroline," Johnson continued, "I want you to call me 'Uncle Lyndon' whenever you see me. Will you remember to do that?" Caroline looked confused, and after Johnson walked out of the room she asked Lincoln, "Is he really my uncle?"

During a 1961 trip to Southeast Asia, Johnson famously invited a Pakistani camel driver named Bashir Ahmad to visit him in Texas and to meet the president at the White House. He told reporters, "What we need on our side are the camel drivers of the world." What followed was a media spectacle. Johnson flew to New York to welcome Ahmad and then flew with him to Texas, where he hosted a barbecue lunch. In Washington, Lady Bird introduced Ahmad to President Kennedy in the Oval Office. Political cartoonists lampooned the visit, but it was largely considered a diplomatic success.

In August 1961, Johnson was given a key assignment when Kennedy sent him to Berlin after the communists built the Berlin Wall to separate East and West Berlin. The Wall was to become a defining symbol of the Cold War. Johnson's job was to voice American solidarity with the people of West Berlin. Hundreds of thousands of people gathered at City Hall in West Berlin to hear him speak. He told the crowd that Kennedy had sent him to communicate the same commitment that "our ancestors pledged in forming the United States: our lives, our fortunes, and our sacred honor." It was Johnson's most important trip as vice president, but during much of the transatlantic flight aboard Air Force Two, reporter Al Spivak recalled how frustrated and broken Johnson seemed. "On virtually the entire flight over the Atlantic he [Johnson] talked the whole time," Spivak said. "The main topic of the conversation was that Kennedy wasn't using him to promote his legislative program."

Johnson was a dutiful vice president and did what was asked of him, but he did not hide his resentment well. His nemesis, Kennedy's brother Bobby, was attorney general and much closer to the president than Johnson was. Adam Frankel, who was a speechwriter for President Obama and worked with Kennedy aide Ted Sorensen on his memoir *Counselor*, recalled attending a dinner decades later at the Kennedy Library where the former diplomat and Kennedy friend John Kenneth Galbraith paid tribute to Sorensen. Galbraith joked, "We all knew who the second most powerful man in the White House was . . ." But before he could utter the words, Sorensen himself interjected and shouted, "Lyndon, of course!"

Everyone at the table had a good laugh. The truth, and everyone knew it, was really Bobby Kennedy.

––––––––

After Kennedy's assassination, Johnson went without a vice president for the remainder of Kennedy's term, which was well over a year. Before ratification of the Twenty-fifth Amendment there was no constitutional mechanism for filling a vacancy in the vice presidency—the next in line to the presidency was the Speaker of the House, at the time John W. McCormack of Massachusetts. When he ran for president in 1964, Johnson picked Hubert Humphrey to be his running mate. Humphrey was Johnson's protégé when both men were in the Senate, and when Johnson was majority leader Humphrey was his conduit to the liberal wing of the Democratic Party. Johnson picked Humphrey, who was a well-respected Minnesotan, to help balance the ticket both ideologically and geographically.

The Johnson-Humphrey relationship has been described as one of domination-subordination. Their solid working relationship on Capitol Hill quickly deteriorated and became a disastrous one in the White House. Outwardly Humphrey seemed busy, overseeing the space program and acting as the administration's go-between to local governments, but within the walls of the White House he was miserable. Johnson adviser Joe Califano said his boss was "essentially a manic depressive, up and down," and he regularly berated Humphrey in front of others. Because Johnson had suffered as Kennedy's vice president, he intended to make sure Humphrey did the same. He insisted that Humphrey stay out of the headlines, leaving more room for himself, and issued a decree that no member of the national press could accompany Humphrey on out-of-town trips. Johnson was obsessive about leaks, and when there was one the first suspect was always the vice president. "Goddammit, Hubert, can't you keep your mouth shut?" he would bellow. "Every time I say something I find your friends writing stories about it."

Humphrey revealed the humiliating subservience that so often comes with the vice presidency in a candid interview with *Fortune*. The reporter asked him what had changed in his life since becoming vice president and he replied, "Don't let anyone ever tell you that it's just a change of office. It's a change of lifestyle, even a change of attitude. For one who is naturally effusive and gregarious and outgoing, it surely require[s] a great deal of self-discipline . . . I've changed. I've become more prudent. I've become more tolerant, too."

Tolerant was the least of it. Humphrey was a self-proclaimed "jolly Santa" and struggled to comply with Johnson's orders at times. When he let it slip in a talk to a group of labor leaders that the administration was planning to ask Congress for an increase in the minimum wage, the news was picked up in the newspapers. Johnson fumed, "I see by the papers where I have a minimum wage program." Humphrey had to ask Johnson for permission to use an official plane, and he had to clear the text of all his prepared speeches—invariably, West Wing aides stripped them of anything meaningful. He was truly in an impossible position working for a larger-than-life president.

Johnson was eventually consumed by the quagmire of the Vietnam War, with protestors gathering every day in Lafayette Park across the street from the White House and chanting, "Hey, hey, LBJ, how many kids did you kill today?" Their cries could be heard in the private residence, and their pleas grew louder and louder each month. Humphrey publicly supported the government's war policy and was tainted by his association with the administration. Privately, however, Humphrey had warned Johnson against military escalation in Vietnam, and for a year he was shut out of any decision-making. Humphrey was torn between personal loyalty to the president he served and his own conscience. "I did not become vice president with Lyndon Johnson to cause him trouble," he said.

Johnson shocked the country when he announced he would not seek the presidency in 1968, leaving room for Humphrey to run instead. But Johnson did not want his vice president to win, in part because he knew that he had turned against the war and would not continue the administration's approach to Vietnam. In a call to Senate Minority Leader Everett Dirksen, a Republican, just days before the 1968 election, Johnson said, "I've told Nixon every bit as much, if not more, than Humphrey knows. I've given Humphrey not one thing, and up to now, Nixon and the Republicans have supported me just as well as the Democrats." He had even convinced Republican governor Nelson Rockefeller of New York to run, hoping that he would beat Nixon and Humphrey. "He told me he could not sleep at night if Nixon was president, and he wasn't all that sure about Hubert [Humphrey] either," recalled Rockefeller, who would later find himself becoming vice president in the wake of Nixon's collapse and disgrace. Five weeks before the 1968 election, Humphrey broke with Johnson and called for a halt to the bombing of North Vietnam in a nationally televised campaign speech. His poll numbers shot up, but it was too late. Just days before the election, Johnson called Humphrey to tell him he had decided to halt bombing of North Vietnam but he wanted to make sure Humphrey did not take too much credit. "If I were you," Johnson instructed, "I would let the laurels come to me, but I certainly wouldn't crow about it." Humphrey was punished for being Johnson's vice president and ultimately lost the election to Nixon.

Jimmy Carter recalled inviting Humphrey to Camp David one weekend when he was president, years after Humphrey left office. It was the very first time he had ever been there. "Poor Humphrey, my friend, they just ruined him there," Carter's vice president Walter Mondale sighed. "He had no status . . . Johnson had this sort of dark, mean side to him. He'd bite."

VIII

Confusion, Conflict, and Musical Chairs: The Rocky Road of Agnew/Nixon, Ford/Nixon, and Rockefeller/Ford

When Ford became vice president, I don't think we thought that Nixon was doomed. I think we thought that he'd been badly wounded.

—GERALD FORD'S SPEECHWRITER ROBERT HARTMANN

After Richard Nixon won the nomination at the Republican National Convention in Miami Beach in August 1968, Donald Rumsfeld, who was then an Illinois congressman, was summoned to Nixon's penthouse suite at the Hilton Plaza hotel in the middle of the night. There he found a room packed with the men who were closest to Nixon, including John Mitchell, who would become attorney general; Senator Barry Goldwater, who ran against Lyndon Johnson in 1964; and the Reverend Billy Graham. The question at hand was who Nixon should pick as his running mate. "What about Mark Hatfield [a Republican senator from Oregon]?" Graham asked. "He's liberal but he's a Christian and I think it would sell in the south."

Nixon responded immediately: "I don't want anyone from the far right or the far left." In the end Nixon's choice, Maryland governor Spiro Agnew, came as a surprise to Rumsfeld and most people in the country, who had never heard of Agnew. Nixon had a few major criteria for choosing a vice president: he must be fully qualified to take over the presidency; he must share similar philosophical and political views; and he must be loyal. Nixon, eight years removed from his demeaning duty under Eisenhower, said, "The president and the vice president need not be personal friends but they must under no circumstances be personally incompatible."

Agnew beat out other much more experienced candidates, including Tennessee senator Howard Baker. Nixon told reporters, referring to himself in the third person: "There is a mysticism about men. There is a quiet confidence. You look a man in the eye and you know he's got it—brains. This guy has got it. If he doesn't, Nixon has made a bum choice." By all accounts, he made a very "bum choice."

"Nixon made a terrible mistake, he was a perfectly unacceptable nominee," Rumsfeld said of Agnew. Not only was he ethically challenged, but he was also lazy. Nixon made Agnew chairman of the Desegregation of the South Committee (a Cabinet committee to manage the transition to desegregated schools), but Agnew, Rumsfeld said, never showed up for any meetings. "Finally, those of us involved got George Shultz to replace him," Rumsfeld recalled, referring to Nixon's secretary of labor. Rumsfeld remembered Agnew being incredibly vain and preoccupied with how he looked, constantly smoothing out the creases of his suit pants in the meetings he did choose to attend. "He was an unusual person, he was interested in his clothes," Rumsfeld said. "I never saw him very interested in substance." Dick Cheney, who worked for Nixon, described the relationship between Agnew and Nixon as a "train wreck."

Like the president he served, Agnew is best known for being forced from office when he resigned in 1973 and pleaded no contest to a charge of federal income tax evasion. He was the first and, so far the only, vice president to ever resign in disgrace. But before then he was Nixon's bullet-headed, well-dressed hatchet man.

Spiro Theodore Agnew was born in Baltimore, Maryland, in 1918, the son of a Greek immigrant whose family name originally was Anagnostopoulos. He worked as a grocery store clerk and at an insurance company during the day and studied law at the University of Baltimore at night. He served in western Europe during World War II and was awarded a Bronze Star. After the war he completed his law degree and set up his own practice outside of Baltimore. He switched his registration from Democrat to Republican and in 1962 ran for county executive and won. In 1966 he was elected governor of Maryland. After Dr. Martin Luther King Jr.'s assassination in the spring of 1968, riots engulfed Baltimore and Agnew took a hard line against protestors and African American leaders. Agnew's response to the rioting helped his bona fides with conservatives and was a major reason why Nixon picked him, hoping to win over white Southerners. But outside of Maryland, Agnew was an unknown. When Nixon announced Agnew as his running mate, headlines asked "SPIRO WHO?"

Though Agnew was the first vice president with an office in the West Wing, he soon learned that he would not be permitted to exert much power. But at the very beginning of his vice presidency, he enthusiastically embraced his role as president of the Senate and spent time each morning with the Senate parliamentarian, Floyd Riddick, to learn parliamentary procedures. He knew he would have to administer the oath to new senators and break tie votes. He diligently studied the names and faces of senators, but his ego was quickly deflated when, after preparing for a speech to the Senate, he was told by Majority Leader Mike Mansfield that he would be given half the time he thought he would have. Agnew considered it an insult, but he would soon learn that it was merely part of the job. He did not fare much better in the White House, where he was put in his place routinely by Nixon's chief of staff H. R. "Bob" Haldeman. "The President does not like you to take an opposite view at a Cabinet meeting, or say anything that can be construed to be mildly not in accord with his thinking," Haldeman told him.

Agnew did not agree with Nixon's groundbreaking efforts to normalize relations with Communist China but felt powerless to stop them. As he began to understand the constraints of the vice presidency he retreated: "A little over a week ago, I took a rather unusual step for a Vice President," Agnew complained in his memoir. "I said something."

What bit of power and influence he was able to cobble together came from his famous attacks against the media, which he singled out as the enemy during nationwide protests against the Vietnam War. He was a strident critic of the protestors and he was the administration's main attack dog, allowing Nixon to stay above the fray, and earning him the nickname "Nixon's Nixon." Agnew enjoyed his adversarial relationship with the press, famously and contemptuously referring to reporters as "nattering nabobs of negativism" and "impudent snobs." Critics of Nixon's Vietnam War policy were "pusillanimous pussyfooters." The creative alliterative putdowns, many of which were penned by Nixon speechwriter William Safire, who went on to become a Pulitzer Prize–winning political columnist, earned Agnew praise in the West Wing. One Nixon adviser glowingly referred to Agnew as "the Robespierre of the Great Silent Majority."

But Agnew's popularity among the conservative wing of the Republican Party soon began to threaten Nixon, and the president never wanted Agnew to forget his place. Aides worried that Agnew's war with the press was going too far and creating enemies for Nixon. "If [Walter] Cronkite and [David] Brinkley et al. should figure they are in a fight for their lives, we would have our hands full with a war we don't need," an aide wrote in one 1969 memo. In a 1972 memo issuing "talking points" to Agnew, the first note is underlined: "No attacks on press *at all,*" and included a forceful reminder to put his own obvious presidential ambitions aside: "No discussion or comment on '76."

When Agnew became embroiled in the corruption scandal that led to his resignation, he turned into a pariah. Two years before Ag-

new resigned Nixon and his top aides John Ehrlichman and Haldeman talked about pushing him out. "Being ahead of the power
curve, as you are at the moment," Ehrlichman told the president, "is
the time for him to resign. I don't know how you do it. I don't know
what the inducement to him is or how you engineer it, but I just see
him as a liability from here forward." As the three men talked Nixon
fumed at press reports of Agnew's golfing on foreign trips. His vice
president is "not over there on a goddamn vacation," he said. "Jesus
Christ, you know, when I went on these trips with my wife, we
worked our butts off, and it made an impression."

Agnew was routinely refused one-on-one meetings with the
president. When he was asked to help defend Nixon after the Watergate break-in, he said he would under one condition: he wanted
a meeting alone with the president. Nixon would not agree to it.
Charges that Agnew took thousands of dollars in bribes from con-
tractors while he was governor of Maryland were front-page news
and particularly harmful as Watergate dragged on. Nixon's aides,
not Nixon himself (like most presidents, he had a deep aversion to
personal confrontation), eventually put an end to Agnew's misery.
Bryce Harlow and Al Haig went to Agnew's office late one night
in 1973 with a clear purpose.

"This is a national crisis," Harlow said. "Congress will undoubt-
edly act. You will be impeached."

After a long pause Agnew said, "What are you here to tell me?
What do you want?"

"We think you should resign," Haig replied.

"Resign? Without even having a chance to talk to the President?"

"Yes, resign immediately," Haig said. "This case is so serious there
is no other way it can be resolved."

Agnew was defiant and refused to resign for several more weeks.
He wrote in his memoir, "Without even an opportunity to be
heard in my own defense, I was to be jettisoned, a political weight
too heavy to allow the presidential plane—now laboring on its last
engine—to remain airborne." The day before Agnew resigned,

Nixon met with him alone in the early evening. Agnew finally got his one-on-one meeting, but it was not under the circumstances he had hoped for. Nixon told him he needed to leave office. Now. Agnew was devastated and wrote in his memoir: "I had become a nonperson. The Vice-President, who had shared the tremendous victory in the national election less than a year before, was suddenly hurled into outer darkness, into the limbo of forgotten men."

Nixon needed to find a new vice president. On October 10, 1973, he summoned Michigan congressman and House Minority Leader Gerald Ford to the White House for a meeting. "Sit down," Nixon told him.

"How serious is it?" Ford asked. "I don't know the details. I only know what I've read in the papers."

During a two-hour conversation Nixon sat, smoking a pipe, sizing up Ford. "Agnew is in trouble, real trouble," he told him. By the time Ford was back on the Hill, Agnew had resigned.

Weeks later, Nixon sent a letter to Agnew's home address in Kenwood, Maryland: "The chair you occupied across from mine at the Cabinet table is, to me, a symbol of the strength and wisdom you brought to that task as well as to the highest councils of the Government itself." But Agnew and Nixon never spoke again. In his memoir, aptly named *Go Quietly . . . Or Else*, Agnew wrote that Nixon "played me as a pawn in the desperate game for his survival" and "I believe he had an inherent distrust of anyone who had an independent political identity."

When Nixon died in 1994, his daughters, Tricia and Julie, invited Agnew to the funeral at his presidential library in Nixon's hometown of Yorba Linda, California. At first Agnew refused, but he eventually relented. He wrote to a friend that he would go even though, he said, "Nixon was an asshole."

———

Gerald Ford was a well-liked senior member of Congress who represented Michigan's very Republican fifth district. He served a quarter century in the House, where he was the minority leader from 1965 to 1973. He and Nixon were very different in temperament—Nixon was moody and formal and Ford was affable and beloved by his colleagues in Congress. But the two men had a lot in common: they were born six months apart in 1913, they grew up in small towns with very little money, they served in the navy during World War II, and they graduated from law school. Nixon was elected to the House in 1946, Ford was elected two years later, and they shared the same political philosophy. They sometimes carpooled from northern Virginia to Capitol Hill together and their wives, Pat and Betty, became friends. Both women were accustomed to their husbands' long absences on the campaign trail. In 1972 Ford went to some two hundred campaign events for Republicans around the country and was away from home 258 days. He and his wife, Betty, were raising four children together, and his frequent absences were taking a toll on her health. Ford agreed he would run once more in 1974 and then retire. He had it all planned out—he thought.

On August 4, 1973, Secretary of Defense Melvin Laird traveled with Ford and about twenty other House members to Groton, Connecticut, to attend the laying of the keel for a nuclear submarine. After the ceremony and luncheon, on the flight back to Washington, Laird took a seat next to Ford and swore Ford to secrecy before he spoke. Watergate was consuming the White House, but Agnew's troubles were also becoming an issue that needed to be dealt with. "We got a problem. It's going to break. It involves the Vice President." Laird added ominously, "You think things are bad now. They're going to get worse."

After Agnew's resignation Nixon became the first president to use the Twenty-fifth Amendment to the Constitution (which was ratified in February 1967, just over three years after President Kennedy's assassination) to choose a vice president. Nixon made the decision as he made many decisions, alone at Camp David. Ford was on a list with three others, all of whom were current or former governors: Ronald Reagan, who was governor of California; Nelson Rockefeller, who was governor of New York; and John Connally, who had been governor of Texas and who served as Nixon's treasury secretary. Ford was Nixon's "oldest and closest friend" of the four, and his ability to be quickly confirmed by Congress "gave him an edge which the others could not match," the president later said. Nixon knew that when he picked Ford there was at least a 50 percent chance that Ford could become president because Watergate was threatening to destroy his presidency.

Gerald Rudolph Ford Jr. was born Leslie Lynch King Jr. in 1913 in Omaha, Nebraska. His father was abusive and his parents separated when he was a baby. His mother moved to Grand Rapids, Michigan, where she married Gerald R. Ford, a salesman who raised Ford and made him his namesake (Ford's name was legally changed in 1935). He played football at the University of Michigan and graduated from Yale Law School near the top of his class. He served in the navy in World War II, and when he returned to Michigan he ran and won a seat in the House in 1948; he was re-elected twelve times. He was incredibly down-to-earth. When he was vice president–designate, waiting to be confirmed, Ford went to a luncheon in Cedar Springs, Michigan, with a group of women who had supported him in his earliest campaigns for Congress. "He stood in line, got a paper-plate lunch from the good people of Cedar Springs," recalled Ford aide Robert Hartmann. "He had to fish in his pocket and pay for it; it was 75 cents, but he didn't have any money." Hartmann loaned him some.

Although Nixon picked Ford in part because he thought he was easily confirmable—the Twenty-fifth Amendment required confirmation by both houses of Congress—Watergate and partisanship took hold. A group of liberal Democratic members of the House thought that if they could stall Ford's confirmation hearings and impeach Nixon, Speaker of the House Carl Albert, a Democrat, who was then next in the line of succession, would become president. They used parliamentary tactics to delay and divert the proceedings. "We had books, loose-leaf books a yard thick of answers to questions that they were going to ask," Hartmann recalled. There were about fifteen hundred pages of raw material the House was reviewing. The hearings got ugly. It was pointed out that the IRS said Ford owed a few hundred dollars because of two summer suits he bought and wrote off as a business expense when he was chairman of the Republican National Convention. But their efforts were unsuccessful. The final votes to confirm Ford were 92–3 in the Senate and 387–35 in the House. He took the oath of office on December 6, 1973.

Nixon took great pains to put Ford in his place, as he had done to Agnew. He even told Ford that in 1976, when it would be possible for Ford to run for president, he was planning to support Texas governor John Connally instead. "Connally was the only man I'd ever seen Nixon around who he looked up to," said Rumsfeld, who was a Nixon aide. "Nixon was just taken by him." Ford, it seemed, was almost relieved. "That doesn't bother me at all," Ford told Nixon. "I think John Connally would be a first-class candidate."

As vice president, Ford walked a fine line serving a president who was under investigation and working under a cloud of deep suspicion. "I was always very hesitant to get involved in any public discussions of the charges [against Nixon] because it could have been, and I think it would have been, misconstrued by people that I was trying to undercut him so I would get the office," Ford said later. But Ford and his staff did not think that Nixon would resign. As Hartmann put it, "When Ford became vice president, I don't think we thought that Nixon was doomed. I think we thought that he'd been badly wounded." Nixon had won reelection in 1972 in a landslide, and Ford assumed that Nixon would finish his term and back Connally in 1976.

As with his selection of Agnew five years earlier, part of Nixon's reasoning for selecting Ford was that he did not think he posed a real threat—he had never run for national office and had no interest in being president. Ronald Reagan or Nelson Rockefeller might have outshined him, he reasoned, but not Gerald Ford. Moreover, Ford's lack of gravitas, Nixon thought, might make his own impeachment less likely. While he was competent, Ford was not considered presidential material. Nixon was looking for a "silent, loyal party supporter," said William Seidman, a Ford aide. "A confirmable guy who was center of the road and was not known for his speaking ability or his brilliance in one particular area." Seidman said, "We were told not to talk with the White House. Really the only ones who talked to the White House were [Philip Buchen, Ford's counsel] and Hartmann, and they did very little of it. I mean it was almost like zero communication." Ford's vice presidential staff had to eat in a separate cafeteria because Nixon's West Wing aides did not want them in the White House mess, composed of three small dining rooms run by the U.S. Navy and located next to the Situation Room in the White House basement. The only person they'd let in was Hartmann, and he said if no one else was allowed then he would not go either.

Ford's home phone number at 514 Crown View Drive in Alexandria, Virginia, was still listed in the phone book when he was vice president (he only canceled it when he assumed the presidency). As Watergate raged on, Nixon's aides were always asking Ford to defend the president in the press, and it was easier to decline when he was not in Washington. "I will have to confess it was easier to be out of Washington than to be in," Ford said years after leaving office. "If you were in Washington, you were pestered hour after hour for some comment on this revelation or that exposure."

Ford felt more comfortable with his former colleagues on the Hill and campaigning around the country for Republicans running for reelection. It was difficult to maintain his personal integrity without looking like he was trying to undercut his boss. "I was really not much involved in the legislative program or policy area," Ford recalled. "I think the White House by design kept me out of it, and to be honest with you, I really was glad to do so.

Because the minute I got involved with Haldeman or Ehrlichman or any of those people, I was uncomfortable." Nixon's secretary of state Henry Kissinger and aide Brent Scowcroft briefed Ford for a couple of hours once a week. "He [Kissinger] and Brent were about the only people in the White House that paid much attention to me," Ford recalled.

Less than a year after being confirmed as vice president, Ford became president when he took the oath of office on August 9, 1974, three minutes after he and his wife, Betty, escorted the Nixons to their waiting helicopter on the South Lawn. It marked the third time in a little more than ten years that a sitting or former vice president became president. Ford recalled the moment, just the day before, when Nixon told him he was resigning. "There was no gushing, there was no dramatic embrace," he said. "There was just the recognition on the part of both of us that the time had come, and his choice had been made and my fate had been decided." Ford's aides felt it was important that Ford's old congressional colleagues stop calling him "Jerry" when they came to the White House, so they made a point of telling them things had changed, that Ford would now be called "Mr. President."

"It was part of trying to restore the presidency and people's loss of confidence in the presidency," Dick Cheney, a top Ford aide, said. "He was Mr. President the day he put his hand on the Bible."

The Fords' youngest son, Steve, was about to begin his freshman year at Duke University when his father suddenly assumed the presidency after less than a year of being vice president. "As an eighteen-year-old kid all of a sudden we all got ten Secret Service agents and life changed," he said. "Trust me, at eighteen that's not really the group you're hoping to hang out with." The Secret Service had to track down the Fords' other sons—Mike, who was newly married and attending a theological school in Boston, and Jack, who was a park ranger at Yellowstone. Jack was out on patrol and got a radio communication to get back to camp. A friend of the Fords arranged for a private plane to fly Jack to Salt Lake City, where he caught a commercial flight to Washington for his father's swearing-in. (Their daughter, Susan, was still in high school and lived with them in their home on Crown View Drive.)

Because the Observatory was not opened as the official vice presidential residence until 1975, the Fords remained in their modest, four-bedroom colonial house for ten days after Ford was sworn in as president so that the Nixons would have time to move their things out of the White House. Bulletproof glass was installed in the Fords' master bedroom, steel rods were put in underneath the driveway to support the heavy armored presidential limousine, and the press crowded the suburban neighborhood. "Our poor neighbors went through hell," Ford recalled.

The new president commuted every day on I-395 for the fifteen-minute ride north to the White House. According to a *Time* magazine story published shortly after he took office, "Many motorists waved a cheerful if somewhat bemused good-morning as the Chief Executive, immersed in his morning newspapers, sailed past them in the lane reserved for buses and car pools." Betty Ford was now first lady, but she was still making dinner. One night, as she was standing over the stove, she turned to her husband and said, "Jerry, something's wrong here, you just became president of the United States and I'm still cooking." When the family moved into the White House they were much more casual than the Nixons had been. Steve Ford would park his yellow Jeep on the circular driveway on the South Lawn, but the Jeep would always be moved by a member of the residence staff to a less conspicuous spot. "It was a running joke that my Jeep didn't look proper out in front of the White House."

When he assumed the presidency, Ford told chief of staff Al Haig to make sure that the Oval Office was swept and that all of Nixon's recording devices were removed. Donald Rumsfeld, who quickly replaced Haig, double-checked. When Ford named Rumsfeld as his chief of staff, Rumsfeld presented him with a schedule for the resignations or retirements of "the people that we felt had to go." Rumsfeld said he had "a precise day-by-day, month-by-month schedule in that regard." By December 31, 1974, the ones that needed to go were gone.

––––––––––

Nelson Rockefeller, a four-term governor of New York and a descendant of the country's most famous wealthy family, was sworn in as Ford's vice president on December 19, 1974, the second person appointed to that office. And he is also the only modern vice president to be dumped from the ticket. (There was a time, when vice presidents were decidedly less powerful, when they were dumped quite often: Abraham Lincoln dumped Hannibal Hamlin for Andrew Johnson in 1864; FDR dumped John Nance Garner in 1940 and Henry Wallace in 1944.) Apart from his humiliating dismissal, Rockefeller is perhaps best remembered for the mysterious disappearance of his son Michael, who vanished off the coast of New Guinea in 1961 and is thought to have been killed by cannibals, and the decidedly undignified way in which he died in 1979, two years after leaving office, alone and with a much younger woman who was rumored to be his mistress.

Nelson Aldrich Rockefeller was born in 1908. On his father's side he was the grandson of the nation's wealthiest man, John D. Rockefeller Sr., the founder of Standard Oil and the world's most famous robber baron; on his mother's side he was the grandson of Rhode Island senator Nelson Aldrich, who had been the Senate's most powerful member at the turn of the century. Rockefeller was one of five boys, and, like most vice presidents, he had always wanted to be president. He was dyslexic but graduated Phi Beta Kappa from Dartmouth College in 1930. He worked in the family business and was a passionate art collector and served as treasurer of the Museum of Modern Art, eventually becoming the president of the museum in 1939. A moderate Republican, he worked in FDR's and Eisenhower's administrations. In 1958, he beat incumbent Averell Harriman to be elected governor of New York by running as a man of the people—ironic given his great wealth. Gregarious and with seemingly boundless energy, Rockefeller sought the Republican nomination for president in 1960, 1964, and in 1968. When asked how long he had wanted to be president, he replied, "Ever since I was a kid, after all, when you think of what I had, what else is there to aspire to?" But the presidency eluded him. In 1964 he lost to Barry Goldwater, in part because of his scandalous affair with the much younger Happy, who became his second wife, and because he was considered too liberal. He campaigned for Nixon in 1968 and 1972, and Nixon appointed him to a government board overseeing CIA activities.

After Nixon resigned the presidency in 1974, George H. W. Bush and Rockefeller were on Ford's short list of vice presidential prospects. He picked Rockefeller, making the announcement eleven days after Nixon's resignation, in part because he thought it was important to have someone from a different part of the country and because he thought Rockefeller could make a good president. Rockefeller had been offered that position in 1968 by Hubert Humphrey but turned it down insisting he was "just not built for standby equipment." This time, however, he relented because he thought it was his moral obligation. "It was entirely a question of there being a Constitutional crisis and a crisis of confidence on the part of the American people," Rockefeller said.

The first president ever not elected to national office, and nicknamed the "accidental president," Ford got high marks for naming a strong seasoned politician as his number two. "Ford didn't pay a lot of attention to politics," said Rumsfeld. "Gerald Ford was not a political person, he was a public servant." But because of Rockefeller's great wealth and questions about campaign contributions he'd made to government officials, his confirmation hearings dragged on for months. Rockefeller grew nervous that his family's vast assets might be put under a spotlight. He even considered withdrawing his name because he was worried the scrutiny might destroy the family. He and his brothers, it turns out, had left different amounts of money to different children in trust funds and wills, and he was concerned it might come to light and tear the family apart. There was even talk of delaying a vote on his confirmation until a new Congress convened in January. "You just can't do that to the country," Ford pleaded with House Speaker Carl Albert and Senate Majority Leader Mike Mansfield. "You can't do it to Nelson Rockefeller, and you can't do it to me. It's in the national interest that you confirm Rockefeller, and I'm asking you to move as soon as possible." Eventually, on December 10, the Senate confirmed Rockefeller and the House followed suit on December 19.

While Rockefeller was fighting for his confirmation, top Ford aides Dick Cheney and Donald Rumsfeld were solidifying power in the West Wing. Ford originally promised Rockefeller chairmanship of the Domestic Policy Council, but he was cut off at every turn and the council's budget was eventually cut. He and Ford met once a week, but he was left out of key policy discussions. When Ford proposed spending cuts, Rockefeller lamented, "This is the most important move the president has made, and I wasn't even consulted." As vice president, he complained he had a simple, soul-deadening job: "I go to funerals. I go to earthquakes." He got the message quickly that his main job was to do what he was told. "I am not in a leadership position," he said. "I am supporting the president. He can exert the leadership and I can support him." He listed the creation of a new vice presidential seal as one of his greatest accomplishments in the office.

Rockefeller had a bad relationship with the powerful duo of Cheney and Rumsfeld and it left him vulnerable. Cheney said that Rockefeller's expensive proposals were not in sync with Republican Party principles. "That led to a very hostile relationship between myself and the vice president," he said. "The president maintained a good relationship with the vice president and I was the bad guy, but that was my role." (It is a role Cheney never seemed to mind.) Rockefeller used the Observatory to entertain, even though he didn't live there, and Cheney was always left off the guest list. "Just about everybody in town who was anybody [was invited]—except me," Cheney recalled. "I never got invited to the vice president's residence until Walter Mondale was vice president in the Carter years."

Rockefeller abused his power in the Senate, Rumsfeld said. "I wasn't a fan," he declared. "He had a tendency to be a little bullyish, he was used to having people agree with him, or he hired them, or he rolled over them." Once, as presiding officer, Rockefeller tried to break a filibuster by refusing to recognize two senators and ordering the roll call over their objections. Ford began planning his reelection campaign in the fall of 1975 and Cheney and others started making the argument that Rockefeller represented the New England liberal elite and was too eager to spend taxpayer dollars, and some aides argued that he was too old (he was not yet seventy). Once the charismatic former California governor Ronald Reagan declared his decision to challenge Ford for the Republican nomination, it was clear to Ford that his aides were right: Rockefeller had to go. "I was convinced with Reagan in the race, and given the makeup of the party, there wasn't any way we could win the nomination unless we captured conservative votes," Cheney said. "We would never get those conservative votes if the guy we had running with us on the ticket was Nelson Rockefeller."

Ford asked Cheney to run a search for Rockefeller's replacement. There were about twelve names on the list and the campaign ran a national poll to see which would most strengthen the ticket. On the list was Anne Armstrong, the first woman named to the Cabinet-level position of counselor when Nixon appointed her to the post in 1973 and who later worked as an adviser to Ford ("nothing critical of Anne," Cheney said, "but the country wasn't there, she cost us twelve points"), and the usual suspects John Connally and Howard Baker. Cheney added a question to the poll asking specifically about Reagan; it seemed to him that Reagan was the logical choice. Poll results showed overwhelmingly that Reagan was the best possible pick. Cheney and Republican pollster Bob

Teeter went to make their pitch to Ford, who was at Camp David. "I tried hard," Cheney recalled.

"I don't want to hear it," Ford said, disgusted with Reagan for challenging him for the nomination. "It's not going to happen."

On November 2, 1976, dark horse candidate Jimmy Carter and his running mate, Walter Mondale, narrowly defeated Ford and his running mate, Kansas senator Bob Dole. In the end, Ford deeply regretted dumping Rockefeller, in part because he lost and also because he felt guilty about the decision. Using Rockefeller's nickname, Ford told friends that "Rocky took himself out," but everyone knew he was forced out. "It was the biggest political mistake of my life," Ford later confessed. "And it was one of the few cowardly things I did in my life."

IX

Getting to Know You . . . or Not: Mondale/Carter, Bush/Reagan, and Quayle/Bush

*I know I have to do something about the speaking
thing. I don't know how he does it.*

—GEORGE H. W. BUSH ON RONALD REAGAN

Walter Mondale was having lunch with his friend and adviser Richard Moe and Hubert Humphrey, who had served as Johnson's besieged vice president, in the Senate Dining Room in May 1976. Mondale was weighing Jimmy Carter's offer to be on his list of possible running mates and Humphrey, who was Mondale's close friend and Minnesota mentor, was there for a specific reason. The lunch was an orchestrated attempt by Moe to get Mondale to accept the vice president spot on the ticket. "I tried to persuade him to be interested. And he was not interested. He loved the Senate, he wanted to stick with the Senate," Moe said. "He saw what had happened to Hubert under Johnson, he saw what was then happening to Nelson Rockefeller under Ford. They were very unhappy experiences." At the beginning of lunch Mondale was "kind of hangdog," Moe said. "I don't want to do it," Mondale told Humphrey. "You didn't have a happy experience."

Humphrey made his opinion clear: "If you have the chance to be vice president, you do it. It's the best thing that ever happened to me. I learned more in that time than I learned since. You can get more done down there in one day than you can get done up in the Senate in a year if you care about public policy, and I know you do." Humphrey was getting worked up and said, "Fritz [Mondale's nickname], you'd be a fool not to do this."

"I saw Mondale's eyes open as Humphrey was talking and I murmured under my breath, 'Thank you Hubert,'" Moe recalled. "I'm absolutely convinced that that was his epiphany moment. He came out of there and instructed us to find out everything we could on Carter and on the vice presidency."

———

"As soon as we knew we were getting the nomination [Carter aide] Hamilton Jordan started to put together a plan for vetting the vice president," said Jerry Rafshoon, who later became Carter's communications director. One thing was clear, it couldn't be a Southerner—Carter had been governor of Georgia—and it couldn't be a Washington outsider—Carter had no Washington experience. Among the finalists was Maine senator Ed Muskie and world-famous astronaut John Glenn, who was a senator from Ohio. "I had the printer make campaign buttons and bumper stickers for all six. Once Carter told me in a meeting that it was [Walter] Mondale, the guy was standing outside the door, and I said, "'Print Mondale!'" (The Carters' eight-year-old daughter, Amy, cried over her father's decision not to name Glenn as his running mate. "I wanted an astronaut to be the Vice President," she said.)

After they were elected, Jimmy Carter gathered his West Wing staff and his Cabinet together and made one thing clear: disrespecting his vice president was the surest way to get fired. "I want you to respond to a request from the vice president as if it came from me," he told them. "If I hear any of you mucking around with the vice president, undercutting him, you're outta here." Carter even personally called reporters at their desks and told them to *Kill the story!* when he heard they were about to publish a piece that was unfair to Mondale.

The effectiveness of any vice president depends on his relationship with the president. Mondale and Carter reached an agreement before taking office that changed the vice presidency forever. Dick Cheney and Al Gore, two successors with very different political ideologies, credit Carter and Mondale with modernizing and strengthening the role. The personal relationship forged between the two men is considered the gold standard. Carter describes their years together as a "family environment." "I think the genius of this and the reason for its success was Carter's commitment," Mondale said. "We both had seen how poorly some vice presidents had been treated and underutilized in the past and both felt it was a waste of talent," Carter said. Vice presidents need to have complete access, Carter argued, pointing to the danger of Harry Truman being kept in the dark by FDR about the atomic bomb.

Walter Frederick "Fritz" Mondale was born in Ceylon, Minnesota, in 1928, the son of a local preacher and a music teacher. As a student he campaigned for Hubert Humphrey, also from Minnesota, and the two developed a lifelong friendship. Mondale graduated cum laude from the University of Minnesota and received his law degree from the same school in 1956. He was soon appointed state attorney general, and a high-profile case garnered him national attention. In 1964, when Lyndon Johnson picked Humphrey as his running mate, Mondale was appointed to fill his friend's Senate seat, where he served until he was tapped by Carter to be his running mate in 1976.

Mondale had flirted with running for president himself in 1976. As a member of Congress, he was a vocal critic of Nixon's controversial position on the Vietnam War. For two years he traveled some 200,000 miles and visited thirty states, but he realized that he did not have the stomach for a presidential campaign. He'd had enough nights sleeping in Holiday Inns, he said wryly. But Carter's offer of the vice presidency was hard to refuse, and when he was asked why he decided to accept, which would surely mean more nights in Holidays Inns, he replied, "I've checked and found out they've all been redecorated." The media took to calling Carter, the former governor of Georgia, and his liberal Minnesota running mate "Fritz and Grits."

Mondale helped balance the ticket geographically, and although both he and Carter grew up in different parts of the country, they both hailed from small towns and were both devout Christians. Mondale was considered more liberal than Carter and had better contacts with labor and the more liberal wing of the Democratic Party. In June 1976, when Carter was interviewing vice presidential candidates, he invited Mondale to his home in Plains, Georgia (where he still lives). The two men expressed aligning views about the vice presidency, and both considered it a wasted asset. Mondale did not want to be standby equipment. "I'm in the Senate, I love the Senate, I can help you there," Mondale told Carter. "I want to work out a deal where I'm in the White House helping you, advising you, representing you, and taking on the tough ones that really break the back of a president up on the Hill."

Mondale and his top aides drew up an eleven-page document that formalized his approach to the relationship. Carter signed off on it with no amendments and no deletions. The deal included Mondale's three major requirements: unimpeded access to the president; the same access to classified material as the president; and unimpeded institutional responsibilities. He also wanted a West Wing office, which he got. (The office is just seventeen steps from the Oval, next to the chief of staff's office.) And he wanted a weekly lunch with the president. In the document Mondale tells Carter, "I believe the most important contribution I can make is to serve as a general adviser to you. The biggest single problem of our recent administrations has been the failure of the president to be exposed to independent analysis not conditioned by what it is thought he wants to hear or often what others want him to hear."

But even Mondale, the first truly powerful vice president, sheep-ishly wrote in the memo that "we could of course" cancel any meetings if it did not fit with Carter's schedule.

Mondale made key decisions that kept him in the loop from the very beginning. When faced with the option to run his campaign office out of Washington, Minneapolis, or Atlanta, where Carter and his team were based, he picked Atlanta. He worked as a sort of shuttle service after the election and escorted Cabinet nomi-nees back and forth from Washington to Plains to meet Carter. Mondale was both an adviser and a loyal lieutenant. "I could trust him to give me candid advice on a variety of subjects, but once a policy decision was made I could also trust that he would represent my decision accurately and with enthusiasm," Carter recalled. "I think the key thing is for a vice president to feel that his advice is heeded and that he has unfettered access to give it." Christine Lim-erick, who worked in the White House from 1979 to 2008 as the head housekeeper, said the vice president she saw most often in the private second-floor residence at the White House was Mondale, whom she also saw in the residence with his wife, Joan, several times having dinner with the Carters.

But of course there were disagreements. Mondale sometimes urged Carter to take a more forceful position on issues, and he was particularly upset by the president's so-called malaise speech and told him he did not think he should deliver it. Mondale cringed when, in the nationally televised 1979 address, Carter chastised the Amer-ican people as the nation was seized by double-digit inflation and soaring gasoline prices. "Too many of us now tend to worship self-indulgence and consumption," Carter said from the Oval Office. He decried a "growing disrespect for government" and "fragmen-tation and self-interest" that prevented Americans from responding to the energy crisis sparked by an overreliance on fossil fuels. He said Americans faced a "crisis of confidence." Mondale thought the speech would eventually backfire and it did—it is considered one of the most politically tone-deaf speeches in American history. "There was a good-natured assumption by some people around the president that our political problems were basically emotional problems that the American people had," Mondale said, "so I couldn't go for that." He even considered resigning because of it. But Mondale never made his dissent public. There were times when Mondale put his foot down privately. Jerry Rafshoon, Carter's communications director, recalls a time when he suggested Mondale take on an issue that he thought was too liberal for Carter. But Mondale wanted to be president one day, so he and his staff were insulted at the suggestion that he would consider putting himself on shaky ground with moderate voters.

In 1980 Carter and Mondale lost their reelection bid after one term in office. Four years later Mondale ran for president against Ronald Reagan and lost. His campaign chairman, Joe Trippi, said no matter how much he told Mondale that he needed to distance himself from Carter, he would not do it. "Mondale was very respectful about keeping the confidence of those conversations with Carter private," Trippi recalled. "Even when it was in his best interest as a candidate to divulge when he privately disagreed with the president, it was not in his character to throw Carter under the bus for his own political gain."

————

George H. W. Bush never wanted to look like he was trying to take the spotlight (as if one could) from Ronald Reagan. In pictures and on TV he took great pains to make sure he didn't look like he was trying to get in the camera shot. He told his staff: "Nobody is to start putting tape down on the colonnade to make sure I'm in the picture. If somebody gets caught doing that, they're gone." He was deferential and had a very clear sense of what was proper for a vice president to do. "He honestly took a literal step back," one Bush aide said.

There was a reason for his reticence. Bush almost didn't get to be vice president, which, in retrospect, meant that he almost did not get to be president either. Reagan initially wanted Gerald Ford as his running mate, but when, at the very last moment, it became clear just how deeply flawed and unworkable the power-sharing arrangement with Ford would have been, Bush became Reagan's second-best choice. The two men had run against each other for the Republican presidential nomination in 1980 and had barely spoken since the primaries. Bush stayed in the race until right before the California primary in June, even after being defeated by Reagan in twenty-nine of thirty-three primaries. The partnership between Reagan and Bush is remarkable not because of how it worked during their eight years in office, but because of what it led to—Bush was the first sitting vice president since Martin Van Buren in 1836 to be elected president.

George Herbert Walker Bush was born in 1924 in Milton, Massachusetts, the son of wealthy Connecticut senator Prescott Bush. After graduating from the prestigious Phillips Academy he enlisted in the navy, flew combat missions during World War II, and was shot down in the Pacific by Japanese antiaircraft fire. He returned home, graduated from Yale University, and served two terms as a Texas congressman. He ran for the Senate twice and lost, but his career is marked by several high-level appointments, including ambassador to the United Nations, chairman of the Republican National Committee, chief of the U.S. Liaison Office in the People's Republic of China, and director of the Central Intelligence Agency.

As vice president, Bush had walk-in privileges for any meeting that he wanted to attend in the Oval Office, and he and Reagan had a weekly lunch, but theirs was mostly a respectful working relationship and not a close personal friendship. "He went up to the residence a handful of times, including after the assassination attempt," recalled Tom Collamore, who worked for Bush when he was vice president. "But were they cocktail buddies upstairs in the residence? No, they weren't." Reagan would often go to the residence after work and make calls to Congress and take copious notes about whatever legislative program they were trying to push through. The next morning a copy was always given to Bush's chief of staff so that he would be kept in the loop.

Bush had a phone on his desk with a red button that was a direct line to the Oval Office. A former Bush aide recalled sitting in the vice president's office on more than one occasion when the red button started flashing. "George, I have this interesting fellow down here," Reagan would say, "would you mind coming down and saying hello?" That was Reagan's way of telling Bush to escort a guest who had overstayed their welcome out, and Bush was happy to oblige.

In many ways Bush idolized Reagan, and it was not easy being number two to the man known as "the great communicator." "I know I have to do something about the speaking thing," Bush told Craig Fuller, his chief of staff. "I don't know how he does it," referring to Reagan's seemingly effortless ability to captivate an audience. Reagan tried to help Bush. Once, he picked up a copy of *Time* magazine and offered an impromptu lesson. "Let me show you what you do," he said, and then he began to read an article aloud to Bush as though he were reading a speech to a crowd. Reagan glanced down at the story once or twice but was able to keep eye contact, reading the story as though he had written it himself. Bush could never come close, and he knew it. Fuller began studying Bush's speeches, and over dinner one night told Bush that 10 percent of his speeches were good. "Gosh, thanks," Bush replied sarcastically. They brought in Republican communications guru Roger Ailes and started putting a videographer in the back of the room during speeches so that Ailes could later study the tapes. The end goal for Bush was simple. He knew he needed to improve so that one day he could be president. He was never as comfortable as Reagan in front of a crowd, but over time, and with lots of coaching from Ailes, he got better.

Bush was often out on the road, campaigning for Republican members of Congress and later running for president. He was able to stay mostly above the fray during the Iran-Contra affair because of his travels. He maintained he knew little of the plan to illegally send money from covert arms deals with Iran to buy weapons for the U.S.-backed Contra rebels in Nicaragua. The biggest political scandal since Watergate brought down two high-ranking officials in the Reagan administration but Bush was relatively unscathed. Like his predecessors, Bush sometimes had trouble defining himself apart from the larger-than-life man he worked for. In 1988, at the height of his campaign for president, Bush wanted Reagan to know that he was going to publicly break with the White House, something no vice president wants to do. At the time the administration was in negotiations with Panamanian dictator General Manuel Antonio Noriega, who had been indicted by two grand juries in Florida on charges of drug smuggling. The White House was working on a deal to drop the charges if he would resign. Critics argued that Bush, who had been director of the Central Intelligence Agency when Noriega was rising to power in Panama, should have done more to stop Noriega's influence on drug trafficking. During the campaign, Bush needed to make his opposition clear. Bush said that if he was elected he would not "bargain with drug dealers . . . whether they're on U.S. or foreign soil." Reagan understood that Bush had no choice but to come out against the administration.

Bush and Reagan had genuine respect for one another. Bush delivered Reagan's eulogy in 2004, and said, "As his vice president for eight years, I learned more from Ronald Reagan than from anyone I encountered in all my years of public life. I learned kindness; we all did. I also learned courage; the nation did." In the great tradition of outgoing presidents leaving notes for their successors, Reagan's note to Bush shows a sense of humor and empathy: "Don't let the turkeys get you down," it read.

Dan Quayle is best known for being a gaffe-prone vice president who became the butt of jokes delivered by late-night comedians, but to hear him tell it, his vice presidency was marked by one important factor: the decency of the president he served. Because George H. W. Bush had been vice president for eight years, he was sympathetic to the specific challenges of the sometimes awkward job—unlike Johnson, who used his years in the humbling role as justification for treating his own vice president poorly. "I'm not sure that any other vice president had the kind of experience I had or the kind of working relationship and partnership we had," Quayle said. "And that is because of who George H. W. Bush is."

James Danforth Quayle was born in Indianapolis in 1947. His father was a conservative publisher who owned newspapers across the country and his grandfather was so well connected that Quayle recalled once walking behind him and his golfing partner, who happened to be Dwight Eisenhower. After graduating from De-Pauw University, and knowing that his draft deferment would be over after graduation, Quayle joined the Indiana National Guard. It was a decision countless other men made at the time, but it was viewed by some as a cowardly way to get out of serving in Vietnam. When it came up during the campaign, Bush aides asked Quayle privately if he had any regrets about not going to Vietnam. "I did not know in 1969 that I would be in this room today, I'll confess," he said. After serving in the National Guard, he graduated from Indiana University Law School and started a law practice with his wife, Marilyn. In 1976 he was drafted to run for the U.S. House of Representatives and won, but he was dissatisfied and had his eye on his next move. Colleagues nicknamed him "wet head" because he spent so much time in the House gym that he often came to the floor to vote with his hair still wet after taking a shower.

In 1980 he was elected to the Senate, where he worked alongside Massachusetts Democrat Ted Kennedy to introduce the Quayle-Kennedy bill, a rare bipartisan effort to help create jobs that got his name in the press. When Bush was searching for a running mate, Quayle frequently dropped by Bush's Capitol Hill office and wrote op-eds and delivered more speeches on high-profile topics so that Bush would notice him.

But it shocked everyone when sixty-four-year-old Bush picked forty-one-year-old Quayle, the first baby boomer on a national ticket. During the campaign Bush had a yellow legal pad in his briefcase, and he would take the pad out and ask governors and members of Congress traveling with him, "Who should I pick?" "Flattering, isn't it? That he would care what they thought," said Collamore, who was a top aide to Bush. "On one level it's politics by flattery and on the other hand he picks up little snippets. You still learn something even if you're not fully taking seriously what you're hearing." Bush decided on Quayle because he wanted to appeal to younger voters and Quayle balanced the ticket regionally, hailing from the Midwest, while Bush had ties to Texas and New England. After the disastrous vice presidential debate with Democratic senator Lloyd Bentsen of Texas, when Bentsen attacked Quayle for comparing himself to Kennedy, Bush told Quayle: "That's just so cheap. You did a great job. You hung in there, you were professional." Before he joined the ticket, Quayle had been reelected in 1986 and was among the country's youngest senators. "The world was mine," he recalled wistfully. He soon learned how difficult his life would become.

All the criticism of Quayle had an unexpected upside for Bush's campaign; any attack coming Bush's way was redirected at Quayle and not at Bush. Bush strategist Lee Atwater told Quayle, "Don't worry. You're the best rabbit we got going. Because the press is chasing *you*." Quayle understood Atwater's logic. "I've always had this theory, which I talked to Nixon about," Quayle said. "The press always has to attack either the president or vice president, this is the way it is." As a former vice president, Bush could sympathize with Quayle. He knew that sometimes being the butt of jokes comes with the job description. Barbara Bush's position, an aide to her husband says, was unapologetic: *No whining allowed. You're forty something and you're vice president of the United States, thanks to my husband.*

When they took office, the two met early each morning and had lunch every week. Bush encouraged Quayle to travel and to talk to religious and conservative groups he was not as comfortable with. (Collamore says Bush had to be "dragged kicking and screaming" to meet with evangelical Southern leaders. "He had strong faith but he didn't wear it on his sleeve.") Quayle enlisted a bigger staff than his predecessors in the hope of reintroducing himself to the American people. But nothing worked. Quayle said that he was being unfairly targeted by both the press and by Bush's own advisers. One of Quayle's biggest detractors was Secretary of State James Baker, who had known Bush for more than three decades. Baker called Quayle's chief of staff Bill Kristol early on in the administration and told him: "Look, we don't know each other that well and I know Quayle doesn't like me that much and vice versa, but we have to try and win this thing." Winning "this thing" meant governing successfully. The unmistakable message was clear: *You stay out of my hair; I'll stay out of yours.*

In May 1992, just months before the presidential general election, Quayle delivered a speech on family values and criticized the blockbuster TV show *Murphy Brown* and its main character for "mocking the importance of fathers, by bearing a child alone." Quayle called Bush afterward and warned him, "Be prepared on this thing. The press is in a meltdown about this." Bush replied, "I've never watched *Murphy Brown*." Quayle had not seen the show either, but an aide told him that using the character as an example was the only way to get the speech any attention. The Bushes' daughter Dorothy was divorced and a single mother at the time, and Quayle wanted to make sure Bush did not think he was attacking all single mothers, so they had a long discussion about the remarks. Family values were not a major issue for Bush, but Quayle's remarks brought them front and center. The press mocked Quayle for denigrating women and for being out of touch, and the speech launched a cultural discussion. Again, Quayle unwittingly provided late-night TV hosts with more fodder.

Quayle's famous misspelling of the word "potato" fueled the cartoonish image of him as a young pretty boy unqualified for the job. It is the one thing most people remember about Quayle, and it has been impossible for him to shake off. At a 1992 spelling bee for sixth-grade students at a New Jersey school, Quayle was handed a flash card with an "e" tacked onto "potato." He sat on a stool as a twelve-year-old student spelled "potato" correctly on the blackboard. "Add a little bit to the end there," he said, watching as the student incorrectly added an unnecessary "e." It launched a relentless assault by the media. In a recent interview, well over two decades later, Quayle wanted to make the point that the teacher handed him a flash card with the word misspelled. "You think that was fair?" he asked rhetorically. "Not even close." He likened it to Gerald Ford being unfairly mocked for being clumsy. "We probably didn't do as good a job as we should have pushing back right away, getting the teacher out there to say that she had given him the card," Kristol said. "We were flying back and we didn't think of it. We probably didn't take it quite seriously enough." Bush thought that Quayle paid too high a price in the media for the gaffe.

As an economic recession dragged on and news emerged that Bush suffered heart fibrillations, people began taking a good look at the young vice president and whether he was fit to be president in a crisis. T-shirts were made with the famous Edvard Munch painting of *The Scream*, with the caption: *President Quayle?* According to some polls, Quayle was less popular than Spiro Agnew. Bush seriously considered dumping Quayle from the ticket in 1992. Bush's son, George W., was the biggest advocate for replacing him with then–secretary of defense Dick Cheney. Cheney said he found out years later that the younger Bush had urged his father to replace Quayle with him. "I like Dan, he's a good guy," Cheney said. "I think Bush subsequently was embarrassed by the troubles Dan encountered. When 43 [George W. Bush] said to 41 [George H. W. Bush] that he should replace Quayle, it implied criticism of a major decision he made." In a remarkable twist, Donald Trump, then a Manhattan real estate tycoon, let Bush aide Lee Atwater know he was available to replace Quayle. Bush wrote in his diary at the time that he found the offer "strange and unbelievable." Trump insists that Atwater approached him: "I said, 'I don't know, Lee, you can check it out if you want, but it doesn't sound right,' because at that time, I had no political aspirations." Either way, Quayle's job was on the line.

Quayle's chief of staff, Bill Kristol, who had many friends in the media, was not amused by the speculation and secretly worked against it. During the reelection campaign, Kristol had a conversation with Bush campaign strategist Robert Teeter. "I'm not going to mobilize any of these people against Bush if he decides to drop Quayle. It's his call." But, Kristol added, "if you guys start leaking on the other hand to try and nudge Quayle off, I get a chance to fight back." When Kristol saw stories that were obvious leaks from Bush's team suggesting that Quayle would be knocked off the ticket, he called a friend at the *Washington Post* and the paper ran a story confirming that Quayle would remain on the ticket. "It was a little exaggerated I would say because I had not been in those meetings," Kristol recalled. "They were annoyed about that, but then they couldn't do anything about it." Ultimately, Bush decided against dropping Quayle because he would have looked disloyal and it would have given the impression, as it had when Ford dumped Rockefeller, that he was questioning his own judgment. Bush used to say, "Loyalty is not a character flaw," and the question about Quayle was the ultimate test of loyalty.

Quayle had some successes as vice president, though they were largely overlooked. When the administrator of the Environmental Protection Agency, William Reilly, disagreed with Quayle about an issue and tried to maneuver around him, Quayle took it straight to the president before Reilly could. Bush sided with his vice president. For Quayle it was an important test of a vice president's leverage. "You've got to support me because if you start letting one Cabinet secretary come to you and you side with him, I'm finished," Quayle told the president. "I won't be effective."

On Election Day in 1992, Quayle went home to vote in Indiana and ran into the man he had beaten for Congress sixteen years before. It was a sign that his political life had come full circle. That night, the Bush–Quayle ticket lost to Bill Clinton and Al Gore. Quayle says he does not regret how his political career ended. He is unshakably loyal to Bush, no matter how painful those four years were. "I had a good run, I came close," he sighed. "I didn't get the brass ring."

X

From Friendship to Betrayal:
The Breakup of Al Gore and Bill Clinton

For almost all those eight years the relationship was one between brothers—that may be a cliché and I run the risk of overstatement, but really we became extremely close.

—AL GORE ON HIS RELATIONSHIP WITH BILL CLINTON IN THE WHITE HOUSE

Forgiveness means surrendering your anger.

—AL GORE AT A SEPTEMBER 1998 MEETING WITH CABINET MEMBERS IN THE WHITE HOUSE RESIDENCE AFTER BILL CLINTON ADMITTED HIS AFFAIR WITH MONICA LEWINSKY

He had mixed feelings about running for president. It was a monkey off his back that he lost.

—A FRIEND OF AL GORE'S ON HIS 2000 LOSS

A couple of days after the shocking results of the 2016 presidential election, a reeling Hillary Clinton received a phone call. It was from her husband's former vice president, Al Gore. She and Gore, once rivals for power and influence inside the Clinton White House, were now in a small club of losers. They are among a handful of seemingly cursed presidential candidates who won the popular vote but lost the Electoral College. Gore said he called Clinton to empathize. "She doesn't need any advice from me," he said. "It was commiserating and reaching out to say, in her husband's famous phrase, 'I feel your pain.'"

Gore and Clinton share a peculiar fate: Clinton beat Donald Trump in the popular vote by almost 2.9 million votes and Trump won the Electoral College with 304 votes compared to Clinton's 227 votes. Gore won the popular vote more narrowly, with 50.9 million votes compared to Bush's 50.4 million, but it is still up for debate whether he won the Electoral College. After the Supreme Court decision halting the thirty-six-day recount of votes in Florida, which, if allowed to continue, might have given Gore the lead, Florida went narrowly to George W. Bush, giving him 271 electoral votes to Gore's 266. Clinton and Gore are the only living examples of candidates who won the popular vote and lost the election, but it happened three times before: in 1824 Andrew Jackson won the popular vote and lost the electoral college to John Quincy Adams; in 1876 Samuel Tilden won the popular vote but lost to Rutherford B. Hayes; and in 1888 Grover Cleveland was defeated by Republican Benjamin Harrison, who lost the popular vote but won more votes in the Electoral College.

Ron Klain, who has the distinction of serving as chief of staff for two vice presidents, Al Gore and Joe Biden, described the Clinton call as "nice" and "courteous," but added, "I don't think there's been deep bonding." Gore's 2000 loss has been compared to a death in the family, and for years afterward Gore disappeared from public view. Karenna Gore, the eldest of his four children, called her father's loss "the heartbreak of a lifetime," and aides to Gore say they've argued among themselves which was worse—Gore won fewer popular votes than Hillary Clinton did, but he arguably won the Electoral College in the closest presidential election in the country's history. Gore often says now that the years have softened the blow: "You know the old saying, 'You win some, you lose some'—and then there's that little-known third category." During the 2016 campaign Gore told Carter Eskew, who ran the advertising and messaging team for his 2000 campaign, that he thought

Trump could defeat Clinton. "I don't think he was shocked by his [Trump's] victory," Eskew said. "But he's not sitting around critiquing her campaign." Gore did not attend the 2016 Democratic National Convention, and aides say there is no closure between Gore and Hillary. "Having the second-highest leader in the party in the United States not being there for the convention seems like a signal," said Ann Walker Marchant, who worked on the Hillary Clinton campaign and in the Bill Clinton White House.

Walter Mondale expressed surprise that Al Gore did not make an appearance at the 2016 Democratic convention. "There's something going on there," Mondale mused. "I went to the convention, vice presidents usually do . . . I expected him to be there." Clinton allies were starting to get angry, and Gore's friends were beginning to get embarrassed, that it was taking Gore so long to endorse Hillary. According to hacked emails published by WikiLeaks, top Hillary aide Huma Abedin wrote, "Hard to put on email but there is no love lost in this relationship." Abedin warned campaign chair John Podesta and aide Cheryl Mills that Gore had told the team in 2015 that Hillary would not get his endorsement. He did not endorse her in 2008, when she ran in the primaries, either, and when Mills suggested Gore might change his mind Abedin replied: "No. [It's] bad." Gore finally endorsed her in a July 2016 tweet. The two had a long and complicated history that was not easy to get over.

Albert Arnold Gore Jr. is the St. Albans- and Harvard-educated son of the distinguished longtime Democratic Tennessee senator Albert Gore Sr. and Pauline LaFon Gore, who was a tireless campaigner and one of the first female lawyers to graduate from Vanderbilt University. He was born in 1948, in Washington, D.C., and spent most of his childhood shuttling between an elegant suite atop the Fairfax Hotel along Washington's Embassy Row and his family's large farm in Carthage, Tennessee, where he spent his summers. From childhood he was told the presidency was his destiny. Before he was born his father instructed editors at the *Tennessean* that if he had a son he did not want to see it buried in the paper.

When Gore was born the headline on the front page read, "WELL, MR. GORE, HERE HE IS, ON PAGE 1." Harvard was the only college he applied to. He was close with his only sibling, Nancy, who was ten years older and died of lung cancer in 1984. Gore was a disciplined and competitive student; under his photo in his St. Albans yearbook classmates wrote, "It probably won't be long before Al reaches the top." At Harvard he earned a degree with high honors in government after writing a senior thesis titled "The Impact of Television on the Conduct of the Presidency, 1947–1969."

Even though he opposed the Vietnam War, Gore enlisted in the army and served in Vietnam as a military journalist. When he returned he worked as an investigative reporter at the *Tennessean* and went to Vanderbilt University Law School before dropping out to run for his father's former seat in the House in 1976. He won the next three elections and eventually ran for and won a Senate seat in 1984. In 1988, at what he describes as the "hubristic" age of thirty-nine, he ran for president and lost his party's nomination to Massachusetts governor Michael Dukakis. He was four years younger than John F. Kennedy when he ran. He has spent his entire life trying to fulfill the impossibly high hopes his parents had for him. Weeks before the 2000 election, Gore's mother told him she was going to Nashville to select her election night dress. "I am so proud of you," she said.

————

From the early heady days of the 1992 campaign, when Bill Clinton defied the traditional strategy of balancing the ticket and picked a moderate Southern baby boomer as his running mate and kicked off a bus tour with the image of two young couples on a double date set to a soundtrack provided by Fleetwood Mac, to the devastating 2000 loss, the relationship between Bill Clinton and Al Gore was nothing short of Shakespearean. After the election, and before the Clintons moved into the White House, Gore was among a handful of key advisers, including Hillary, who huddled around a table at the Governor's Mansion in Little Rock and helped pick Cabinet appointees.

But there were signs of trouble to come from the beginning. The bonds that were forged were strong but not unbreakable. They went through a lot together, and the highs were very high and the disappointment incredibly difficult. Unlike Biden and Obama, who grew closer over time, Gore and Clinton's relationship grew more strained as the years went by.

In 1992, when the young couples took off on the campaign bus trip from New York to the heartland, Hillary's close friend and campaign scheduler Susan Thomases tried to get the Gores off the bus. "She was dismissive of then-senator Gore during a conference call, basically calling Al's idea to continue the bus trip 'stupid,'" said Roy Neel, who was Gore's chief of staff in the Senate and later as vice president. "You just don't do that." On that conference call with Clinton aides in the War Room in Little Rock on the very first day of the trip, Gore made it clear that would not be happening. He and Clinton were a team on an almost equal footing, and he would not be subservient. In a *Los Angeles Times* poll conducted less than three months before Clinton announced Gore as his running mate, Clinton came behind then-president George H. W. Bush and Texas billionaire Ross Perot, who ran as an independent. When he picked Gore as his running mate that changed. Gore had an attitude that they won the White House together, and while it was not a 50/50 relationship, he thought he should be given 20 percent of the picks for Cabinet secretaries and 10 percent of the seats at state dinners. His attitude was, *I helped build this, I'm a co-owner, a minority owner, but a co-owner, and so I should get some share of what we're building.*

When a rumor circulated that Thomases was making the case that Hillary should have what had become the traditional vice president's office in the West Wing, Gore and his team were furious and the rumor was quickly quashed. (Hillary eventually settled for another office, though one still in the West Wing.) Gore had

to wait until Clinton picked Mack McLarty as his chief of staff to
come to a formal understanding of what his role would be in the administration. Neel sat down with McLarty and used the Mondale-
Carter memo as their basic understanding and rewrote parts of it.
Gore and Neel were called to a 7:30 p.m. meeting with Clinton to
discuss the memo at the Governor's Mansion. The meeting didn't
begin until 10:00 p.m., with Clinton ticking through each point.
He stopped when he read that Gore's vice presidential chief of staff
would also be an assistant to the president. "This is very impressive,
has this ever happened?" he asked. Neel said that Mondale's chief
of staff had been an assistant to Carter, and Clinton nodded, "Well,
that sounds fine." Gore would have a weekly lunch with Clinton,
and be part of every major national security debate in the White
House. At the first Cabinet meeting the day after the inauguration,
Clinton told his team that a call from Gore should be treated like
a call from Clinton.

Much of the relationship between Clinton and Gore was
negotiated—it was about give-and-take. "Gore regarded it as a
contractual obligation that they have a weekly lunch," said Ron
Klain, who was Gore's chief of staff after Neel left. Clinton's chief
of staff dreaded having to call Klain to tell him when Clinton had
to reschedule. And there was always a negotiation: if Clinton could
not do their weekly lunch one week—maybe because he had to
travel or his schedule was too jammed—he would have to make it
up to Gore and meet with him at eight o'clock at night, or whenever he could find the time. "The meeting happened every single
week no matter what," Klain recalled. "Clinton knew it, his chief
of staff knew it. It *had* to happen." Clinton's staff was annoyed at
having a weekly lunch locked in, but they also admired Gore for
getting so many concessions from the president.

Before Gore's lunches with Clinton, his aides would send
memos to Klain, who religiously collected questions and notes
and handed them to the vice president before lunch. During their
lunches, Gore would update Clinton on the most important issues

he was working on and sometimes act as a conduit, carrying messages, including requests for the president to come to certain events that his West Wing staff had turned down but that Gore thought were important. "Please reconsider" would be the message. Gore would diligently work through his list, and when lunch was over he would return to his office and take out a pair of scissors. Sitting at his desk, he'd cut out different requests or areas of discussion and write a little note with the president's response. Klain would collect the shards of paper and deliver them to the aides who had suggested a question or topic. Gore remembers the lunches as a mutually beneficial ritual. "If the relationship between the two principals is a really good one, those weekly lunches become a kind of oasis for the president to really let his or her hair down and talk about issues and problems of all varieties, including relationship issues, including personal chemistry affecting the interactions of different members of the White House staff and the Cabinet."

But there is a reason why there are former staffers who to this day still identify themselves as "Clinton people" or "Gore people," and much of that is tied up in the complicated relationship between Gore and both Clintons. Gore thought he was the heir apparent and had served as a dutiful vice president for eight years, only to have his presidential ambitions quashed by the president's affair with twenty-two-year-old intern Monica Lewinsky. The affair began in November 1995 and Clinton and Lewinsky had almost a dozen sexual encounters over the next year and a half, most of them in the Oval Office. The affair became public more than two years after it started and the media firestorm consumed the rest of Clinton's presidency. "The Lewinsky revelation basically froze the presidential campaign," Neel said. Gore, Neel recalled, "had to come to the aid and defense of the President at that point. That really knocked us for a loop for a long time." Maybe, some in Gore's camp speculated, Clinton did not want Gore to succeed.

"One of the things about being vice president is that it is infantilizing," said Carter Eskew. "Gore refused to be infantilized and yet in the public's mind he was always the one standing at attention behind Clinton . . . The idea that Al wasn't strong enough and he needed Bill to push him over the finish line would send a very bad message to voters." Even without the challenges presented by the scandals that engulfed the Clinton White House, vice presidents win their party's nomination but are faced with the question: How, after eight years of one presidency, can a sitting vice president convince the American people that there will be change? In American politics, "change" is a very powerful word.

————

Rewind almost a decade. In June 1992 Gore was at the Ashby Inn in rural Virginia, about sixty miles outside of Washington, with his friend Carter Eskew. At the time, Eskew said that Gore "was just a senator from Tennessee and not somebody you would recognize." The waitresses knew Eskew, who was a frequent guest, and asked him who he thought Clinton would pick as his running mate—none of them recognized Gore. "I remember distinctly him saying to me, 'I believe Clinton's going to choose me,'" Eskew recalled. "I was locked in the old-school way of thinking, 'Why in the hell would he pick a clone? Someone who's the same age; neighboring state; new Democrat. I literally thought he was delusional and I said, 'What makes you think that?' And Al's best reason was that in his meetings with Clinton he had really connected with him. Of course everyone felt they connected with Clinton. That was Clinton's great gift. I thought, this poor bastard is really going to be disappointed. And just days later, in the middle of the night, Al called me and he said, 'I've been picked.'"

Gore got the call at his Carthage, Tennessee, home just before midnight on July 8. His nearly three-hour meeting days before with Clinton was exceptionally important. Roy Neel recalled the excitement of the night. Dozens of television satellite trucks and reporters gathered at the bottom of the hill at the Gore family's home. "At one point we have binoculars looking at reporters and seeing that they had binoculars looking at us," Neel said. Early the next morning the Clinton campaign sent a jet to Nashville, and Clinton and Gore began a whirlwind partnership that spanned almost a decade.

But once he made it to the White House as vice president, Gore was in the unenviable position of often facing off against the most powerful first lady in American history. There can only be one second-most-important person in the White House, and there were two strong contenders for that title—and, for the first time since Edith Bolling Galt Wilson took over the presidency after her husband's stroke, it was the first lady and not a member of the president's senior staff challenging the vice president's position.

One of the reasons why Clinton picked Gore was because he reminded him of Hillary: strong, organized, smart, and loyal. But the White House is surprisingly small. The two fought over turf almost like a brother and sister fighting over their father's attention. "It was a team of rivals," said one former Gore aide. Hillary had her own team in the White House, the so-called Hillaryland, a group of mostly female advisers. She was so often surrounded by adoring staffers vying for her attention that Hillary would joke and say "Mommy, am I pretty?" mocking their desperate attempts for her approval.

"The relationship between presidents and their vice presidents is like a marriage," said Steve Ricchetti, who was Bill Clinton's deputy chief of staff and later Joe Biden's chief of staff. "You have to work at it and you have to be intentional about it." And no one could compete against the president's wife. "She was at the table," says Ann Marchant, who worked for Bill Clinton as director of research in the White House. "If anyone was rankled by it they were not there for long!" When asked in an interview for this book what it was like working in an administration with the most active family member since Robert Kennedy, Gore declined to answer and said facetiously he thought the interview might have to be cut short.

The Clinton administration developed a theory it called sequencing, with big initiatives like healthcare reform, welfare reform, and reinventing government competing against each other because the administration wanted to tackle only one large issue at a time. There was a lot of arm wrestling between Gore and Hillary over whether or not reinventing government, which was his assignment, would go before or after healthcare reform, which was hers. Hillary won out.

Within the structure of the White House, they were sometimes pitted against each other because they both had assignments they wanted to succeed at, and there's only so much oxygen to go around. In 1994 Nelson Mandela became South Africa's first black president and his inauguration was an important political event, so important that two of the highest-ranking Clinton officials attended it: Hillary went on one plane with her staff and Al Gore went on another. Who was going to represent the United States at Mandela's inauguration? Neither were heads of state. It struck Gore staffers as strange that they did not go together, as one delegation. Hillary's West Wing office was close to Gore's, but they were rarely seen talking casually with each other. Soon Gore's staff came to the realization that the vice president and the first lady were just too much alike: "Of course they don't like each other, because they *are* each other," said a Gore aide. "They're exactly the same person: wooden on the stump; come off as cold and phony and preachy; nobody likes them because they're the smartest kid in the class. But in fact they were also both very organized, linear, conscientious thinkers." After every campaign stop Gore would walk back on the plane or bus and yell, to no one in particular, "Wires!," and staffers would scurry to get him the latest from the AP and Reuters so he would have the most up-to-date information at his disposal. And Gore sprinkled so many facts into his remarks and in debate prep that Clinton adviser James Carville offered some unsolicited advice: "One fact! One message! One theme! Say it three times!"

Sarah Bianchi, who was a policy adviser to Gore and later to Biden, said of Hillary Clinton and Gore: "They're both very analytical. They don't have the connectivity. Both of them were likely to answer your heartfelt question with a policy answer." Gore's speeches could be so dry that even low-ranking aides felt the need to chime in with advice. During the 2000 campaign, one Clinton aide sent Gore an email suggesting that he keep ceremonial speeches to less than a thousand words and major policy speeches to less than two thousand. "The Gettysburg Address, of course, was about 266 words," the aide pointed out. "One other thing I think your speechwriters ought to keep in mind: perhaps they could avoid the word 'acknowledged' altogether. I think people feel much better about being thanked rather than acknowledged."

No one could have predicted how much Gore's relationship with Bill Clinton would deteriorate, but there were signs at the beginning that set the stage for rivalry. In 1992 most members of Clinton's campaign team opposed the choice of Gore because they thought Gore would never want to play second fiddle. But Clinton ultimately decided on Gore because he had a strong set of strengths to balance his, including foreign policy experience in the Senate. In Gore, Clinton saw a loyal soldier. Bruce Reed was Al Gore's speechwriter during the Clinton-Gore 1992 campaign and worked with Gore on his convention speech. "There was a line in the speech with Gore saying, 'I don't know what it's like to lose a father, but I know what it's like to lose a sister and almost lose a son,' and Clinton in his acceptance speech after that had a line about losing his father," Reed said. "The campaign staff fought and fought and fought to try to get us not to use that line . . . We ended up keeping it." (The line humanized Gore but did draw some criticism for using his son's pain to win political points.) Clinton's campaign staff wanted to save the best material for Clinton; the presidential candidate gets the first choice on everything. Clinton was not pleased with Gore's performance in a fall debate with Vice President Dan Quayle, who said Clinton lacked the moral authority to be president. Gore, Clinton said, should have come to his defense more stridently. Hillary agreed.

During the first term Gore had a good relationship with Clinton, despite his turf battles with Hillary. "For almost all those eight years the relationship was one between brothers—that may be a cliché and I run the risk of overstatement, but really we became extremely close," Gore said. Gore, who is often lampooned for being unbearably dull, actually has a dry sense of humor and used to make Clinton laugh. Like Clinton, Gore's weight was always an issue and he often requested workout equipment in his hotel rooms when he traveled. The navy stewards would put a plate of doughnuts in the middle of the table during Cabinet meetings, and "you couldn't get at the doughnuts for Clinton and Gore stuffing their faces," said Elaine Kamarck, who worked in the White House from 1993 to 1997. "They both were like little porkers who ate too much and then tried to jog to get rid of the pounds."

Gore was good at getting Clinton, who is famously easily distracted and disorganized, to focus. He pushed Clinton to act decisively and swiftly in the Balkans and Haiti. Clinton was wrestling with how to balance the risks and benefits of U.S. intervention in Bosnia when Gore pushed for intervention to respond to the Serbian slaughter of Bosnians in the former Yugoslavia. Gore sat in the Oval Office in the chair on the president's right that's used by visiting heads of state, along with Clinton's secretary of defense, secretary of state, national security adviser, and Gore's national security adviser Leon Fuerth. He told them about a conversation he had with his oldest daughter, Karenna, who was in her early twenties. At breakfast in the Naval Observatory one morning she showed him a front-page story in the newspaper about a woman from Srebrenica—where thousands of Bosnian Muslim men and boys were killed by Bosnian Serbs—who hanged herself after being gang raped. The horrific story stuck with Gore. "Is it really true that the United States cannot do anything about this?" Karenna asked her father. "I recounted the conversation that I had had over breakfast," Gore recalled. "And I said, 'What is the answer to her question?' And our policy changed in that moment and he [President Clinton] directed the team to draw up plans." In the summer of 1995 the United States took a leadership role to stop the war in Bosnia.

For Gore, the more briefings, the better. Aides worried that he was consumed with impressing Clinton, showing up at every meeting and essentially "staffing" the president. But Gore brought a steadiness to the White House that Clinton, who was almost always running hours late and operated on what both political and residence staff referred to as "Clinton standard time," was sorely lacking. When Clinton was waffling on his 1993 economic plan, Gore told him, "You can get with the goddamn program!" Gore aide Elaine Kamarck said, "One of the other tragedies of this White House is that they were a terrific team in the first term." As Gore remembers it, there were few instances when they disagreed. "There were one or two times in the eight years where he had a different view and I was unable to argue him out of it, only one or two or maybe two or three."

Gore's similarities to Clinton both in age and Southern roots made him think that if he just tweaked a few things—became more affable, loosened up a bit—then he, too, could be president. Gore had been training his memory for decades, taking speed-reading courses and reading about methods to help him remember names and faces quickly—something Clinton did effortlessly. All those times Gore had to stand behind Clinton during formal remarks only added to the impression that he was a wooden creature of Washington.

Jamal Simmons, who worked for Gore when he was vice president and was deputy communications director for Gore's 2000 campaign, said there was the feeling inside the Gore campaign that the attention of White House aides was divided between Gore's presidential run and Hillary's run for the U.S. Senate in 2000. Gore and his advisers struggled with whether to take on all the weight of Bill Clinton's presidency and take Clinton, an impeached president embroiled in scandal, up on his offers to campaign for Gore, or to distance themselves from him.

There were also deep problems with Gore as a campaigner. Sarah Bianchi recalled how nervous Gore was campaigning in New Hampshire in 2000 and how he would want to be briefed on exactly who he was talking to at every turn. He was more comfortable talking about global warming or the budget—he needed index cards in case he forgot what he was going to say in a send-off to a loyal aide who was resigning. A staff member remembered how he would stand in front of a voter after a rally and say something stilted like, "So I understand that you're suffering from cancer." Aides would try to loosen him up by telling him jokes before he went onstage. Once when his speechwriter found a funny story to include in one of his speeches, Gore asked him, "Where did you get this anecdote?" Then he pressed him, "Did you get the author's approval?" Gore seemed so uncomfortable reciting the humorous lines that he bungled them. Gore rarely thanked his staff. "Gore is an intellectual," Bianchi said. "You're a staff person, you're part of the process." He is a policy wonk who enjoyed conversations that might start out with, *What was the most impactful thing that's happened in healthcare ever—immunizations or hygiene*? Not a resounding campaign rallying cry.

In the end, Gore's reaction to, and specifically Tipper Gore's disgust with, Clinton's affair and his repeated denials drove Gore away from running on the Clinton record in 2000 and he struggled to find balance between his past and his future. Clinton had told his vice president as he was standing in the entryway of the Oval Office that he did not have sex with Lewinsky. Gore looked unconvinced and spent the following months of 1998 complaining to aides about how difficult the scandal was making it to get policy done, without directly weighing in on his own feelings of personal betrayal. When he asked an adviser one day, with the door to his office closed, whether he believed Clinton was telling the truth, the aide said flatly, "No." Gore was silent. Clinton only came clean about the affair to Gore a week before he admitted the truth to the public and after he had denied the allegations for seven months.

On August 10, 1998, in the back of a limousine in Chicago, Gore finally asked Clinton whether it was true. When Clinton simply expressed how disappointed he was with himself Gore replied: "Well, this is the first time you have apologized to me personally." After his confession, Clinton was in good spirits but Gore emerged from the car upset, offended, and conflicted about what it all would mean for his own presidential ambitions. Seven days later, on August 17, 1998, Clinton became the first president to testify as the subject of a grand jury investigation. After a marathon four-and-a-half-hour testimony conducted via closed-circuit television from the historic Map Room of the White House, Clinton confessed on national television later that evening to the "inappropriate relationship" with Lewinsky. A month later, Clinton invited members of his Cabinet to a meeting in the private residence to apologize for his affair with Lewinsky and for lying to them about it. The president talked about his policy accomplishments and admitted to his bad behavior. He said that he owed them an apology, much as he did his own family, saying he had "work" to do to mend the broken relationships there. Health and Human Services Secretary Donna E. Shalala was less forgiving than some of her colleagues, telling Clinton that he had an obligation to be a moral leader. Clinton told her that if personal morality was so important, then Nixon would have been elected president in 1960 instead of Kennedy, who was a known philanderer. Clinton said he had been furious for four and a half years, when Independent Counsel Kenneth Starr began looking into charges, including the Whitewater land deal and the firing of several longtime White House employees in the Travel Office, a scandal known as Travelgate. The revelation of Clinton's affair with Lewinsky stemmed from Starr's investigation and Clinton considered the whole thing to be a witch hunt. Some Cabinet members quoted Scripture in the meeting, as did Clinton, but Gore had a specific message: Clinton had made mistakes but now must understand that receiving "forgiveness means surrendering your anger."

Clinton's impeachment cast a permanent shadow on their joint accomplishments. After Clinton's grand jury testimony, Gore did not publicly defend him. He released a brief statement that included the sentence, "I am proud of him, not only because he is a friend, but because he is a person who has had the courage to acknowledge mistakes." Privately, Clinton bristled at his vice president's judgment and moralizing. Gore watched part of the House impeachment vote with Clinton in the private study off of the Oval Office. "It's not fair, what they've done to you," Gore told Clinton. But he was deeply worried about the damage Clinton's indiscretions would have on his own future.

———

In an interview with ABC News's Diane Sawyer at the beginning of his 2000 campaign, when he wanted to talk about almost anything else, Sawyer asked Gore what he thought when he heard the truth about Clinton and Lewinsky. "You can imagine how I felt," he said. "What he did was inexcusable, and particularly as a father I felt that it was terribly wrong." Gore was not entirely free from scandal himself. He made multiple fund-raising phone calls from his West Wing office during the 1996 reelection campaign and he was criticized for his aggressive fund-raising role even by some Democrats.

Gore knew what he had to do in order to win, but creating his own identity while celebrating the administration's successes and downplaying the scandals that consumed the Clinton White House would prove too difficult. "You're number two," Gore said, "and whether it's in politics or business or other professions, you have to make a transition from being number two to number one."

Gore pollster Stan Greenberg made the case that Gore should not run on the Clinton-Gore record and should distance himself from Clinton. In a memo ten days before the November general election, Greenberg urged the Gore campaign team to use Gore's running mate, Connecticut senator Joe Lieberman, to do so, largely because of his previous moral denunciations of Clinton: "Lieberman will have to take the lead in attacking Bush not being up to the task of extending our prosperity, but he will need to show that Gore-Lieberman is consumed with restoring the moral force of the family." Gore chose to follow Greenberg's advice. They needed to capture the under-fifty non-college-educated white women who questioned Gore's morality because of his association with Clinton, Greenberg said.

Former Clinton aides maintain that not using Bill Clinton to campaign for him was the single biggest mistake Gore made. But, according to an ABC News/*Washington Post* poll released eight months before the election, half the public believed at the time that Gore was tainted by his association with Clinton. Responding to the statement "Al Gore is too close to Bill Clinton to provide the fresh start the country needs," 49 percent of those polled agreed and 49 percent disagreed. When Gore's campaign was not going well his aides laid the blame on Clinton. "There are only really three possibilities," says Ron Klain. "One is, *It's going quite fine and just tune out all this noise,* which is what I said. The other option is, *It's not going well because you Mr. Vice President are not performing well as a candidate,* and the third is, *It's not going well because this other guy over here is totally screwing you up.* And obviously the third is appealing advice to give to someone. *It's not your fault, it's the other guy's fault.* And so resentment of the president became a big thing and my position was, I would not indulge that." In the end, Klain said, it was tempting for the vice president to believe that none of it was his fault. "Those of us who were telling him no, it's not all Clinton's fault, became unpopular, and one by one we all got fired, Bob Squier [a close friend and adviser to Gore] first, Mark Penn [Gore's pollster] second, and myself third."

Carter Eskew ran the message and advertising team for Gore's presidential campaign and said that the decision not to use Clinton was made for purely political reasons and not emotional ones. But it was impossible to separate the personal from the political.

Although Hillary and Gore were horrified by Bill's affair with Lewinsky, it stuck with Tipper most. "She was furious at him," said Elaine Kamarck. "She was much madder than Al, but he was mad, too." In the end, he was mad because he was betrayed, but Hillary had been through this before. She knew who she was married to—Gore thought they were genuine friends. The election was a fight over undecided voters who made up a very narrow slice of the electorate—half of whom thought Clinton was a pariah. Eskew said the idea of Gore relying on his boss to carry his message would have been foolish and would not only have not helped him but would have arguably hurt him. And there was no way to use Clinton in some states and not in others; it had to be all or nothing. "You can't hide the president of the United States—if he goes out on the road he doesn't just speak to one population, he speaks to all populations."

Clinton thought Gore needed to get over the scandal, as Hillary had. Ten days after his acquittal by the Senate on February 12, 1999, Clinton told his friend historian Taylor Branch that a Gore-Clinton (as in Hillary Clinton) ticket could not be beat in 2000. "Hands down," he said, "but I don't think Al would ever do it." His own wife, Clinton said, had been quicker to forgive him than his vice president. "I thought he was in Neverland," Clinton said of Gore. Clinton was incredulous: Hillary ran on the Clinton-Gore record and went on to win her New York Senate race in 2000 by double digits, marking the first time a first lady had ever been elected to public office. If she could get over it, why couldn't he?

The two men who had once been friends and partners met alone for a two-hour conversation after Gore lost the election, a meeting that was deliberately kept off their public schedules. The discussion was a political autopsy of what went wrong, and there was plenty of blame to go around. Clinton said he understood why Gore wanted to run on his own, but at the end of the day, Clinton argued, he could have been helpful in some key states, particularly his home state of Arkansas, which Gore lost by less than 5 percentage points. In New Hampshire, Clinton said, Bush won by a slim margin, and Clinton had approval ratings over 60 percent there. If he had gone there, Clinton argued, Gore would be president. "I think you made a mistake not to use me more in the last ten days," Clinton told him.

Throughout the tortured five-week period between the November 7 election and the Supreme Court's December 12 ruling, Democrats argued that the vote in Florida was rigged, that Florida governor Jeb Bush, George W. Bush's brother, was unfairly tilting the recount. Gore reluctantly conceded and on December 13, he told the American public that he was giving up his quest for the presidency "for the sake of our unity as a people and the strength of our democracy." Weeks later, when the Bushes moved in on Inauguration Day 2001, it was hard to ignore Gore's obvious annoyance when the incoming president, George W. Bush, walked into the Blue Room as coffee was being served. Bush, in a buoyant mood, said to the butlers, many of whom he knew from his father's four years in the White House: "Told you I'd be back!"

When Gore lost it was understood that the Clintons were in charge. "The Democratic Party was in the hands of the Clinton machine and when Gore lost it meant that the party stayed loyal to Clinton," said Donna Brazile, who was Gore's campaign manager in 2000. "They blamed Gore for not embracing Clinton, the fact that Gore lost was a validation that he should have run with Clinton and not away from Clinton."

Clinton and Gore did not stay in touch in the months following Gore's defeat. But tragedy brought them together. On 9/11 Gore was in Austria and his second call was to Bill Clinton (his first was to Tipper). Gore flew to Toronto and, because flights were canceled, he rented a car and was planning to drive, with an aide, all night to Washington, D.C., where President George W. Bush had scheduled a prayer service at the National Cathedral for the next day. But he got a call from Clinton around 11:00 a.m. "Why don't you stop off in Chappaqua and spend the night with me?" he asked, freshly returned from his own trip. Gore arrived after 3:00 a.m. and the two stayed up talking and drinking all night. "It's hard to say we were having a good time in the midst of that terrible tragedy—we were not—but it was great to be with him again," Gore recalled.

Al Gore and Hillary Clinton share a sense of historical irony, having both lost close contests in part because of their association with Bill Clinton. Since his 2000 defeat, Gore has won the Nobel Peace Prize for his work on climate change and an Academy Award for *An Inconvenient Truth*, his 2007 documentary on global warming. In 2017 he released *An Inconvenient Sequel*. He has become better known for his activism than for his eight years as vice president. He told Klain that he regretted not spending more time in the 2000 campaign talking about climate change and pushing environmental issues. Aides recognized the issue was among the few things that made him come alive on the campaign trail, but, at the time, it had not yet captured the public's attention.

According to a friend of Gore's, losing the election was devastating for him but, in the end, it was a blessing. For a politician who had spent his career positioning himself for the presidency, "it was a relief that he had gotten that far," the friend said. "He had mixed feelings about running for president. It was a monkey off his back that he lost."

XI

The Shadow President: Cheney and His Sidekick Bush

They've leaned on him [Cheney] a bunch, probably Pence the most. They want to know how OVP [the office of the vice president] runs, what's its role and how you structure it.

—CHENEY AIDE NEIL PATEL ON TRUMP ADMINISTRATION
OFFICIALS ASKING HIS OLD BOSS FOR ADVICE

He was marginalized so much that it didn't really matter. He was the grouchy uncle yelling at everyone to get off the grass.

—FORMER AMBASSADOR CHRISTOPHER HILL ON
CHENEY'S ROLE IN BUSH'S SECOND TERM

Without question it was the most tense element of the relationship between myself and the president . . . It's a travesty.

—DICK CHENEY ON GEORGE W. BUSH'S REFUSAL TO PARDON HIS
FRIEND AND FORMER CHIEF OF STAFF SCOOTER LIBBY

Dick Cheney splits his time between his ranch in Wyoming and his elegant home in McLean, Virginia, across the street from Bobby Kennedy's old estate, Hickory Hill, and a stone's throw from CIA headquarters. He hosts the occasional journalist, usually offering cups of coffee and making small talk in his cozy office lined with books and a bust of Winston Churchill that stands in the corner. He does not have a huge security detail, but a couple of aides who worked for his family in the Observatory have left their jobs to work for him now. Mike Pence has asked Cheney, his most recent Republican predecessor, for advice, which he is happy to provide. Pence has an office on the Senate side of Capitol Hill—where vice presidents traditionally have an office and where they technically preside as president of the Senate—and one on the House side, too. Cheney said his example of having two offices on Capitol Hill inspired Pence to do the same.

But Cheney, like most establishment Republicans, is worried about President Trump. "I think Cheney's just as lost as most of us," said a former aide who is still close to Cheney. "I don't think he really gets it and I don't think most Republicans get it. Their voters abandoned them for this crazy guy and I think they're either in denial or they don't want to think about it." Another former senior Cheney aide said wryly, "I love that his legacy looks better and better every day."

As vice president, Cheney rather enjoyed his image as one of the most polarizing figures in recent American history. Cheney's former chief policy adviser Neil Patel said that, during the second term, he bought Cheney a Darth Vader costume complete with the iconic mask, and he wore it into the Oval Office one day and posed for a picture with aides. After leaving the White House, Cheney drove a black pickup truck with a Darth Vader hitch cover.

His relationship with George W. Bush was all business. They were never especially close to begin with—they watched election night returns in 2000 and 2004 separately at the beginning of the night and only gathered together when it looked like they would win. The two did not socialize, but there was more give-and-take than the satirical image of a mastermind pulling the levers of power behind the scenes. Once Cheney joked, "Am I the evil genius in the corner that nobody ever sees come out of his hole? It's a nice way to operate, actually." Former Vice President Dan Quayle went to visit Cheney around Inauguration Day 2001 and delivered a bleak message: "Dick, you know, you're going to be doing a lot of this international traveling, you're going to be doing all this political fund-raising . . . you'll be going to the funerals. I mean, this is what vice presidents do. We've all done it." With a sly grin Cheney replied, "I have a different understanding with the president." Cheney's dark image was a mutually beneficial part of their dynamic: Bush did not mind being underestimated. If Cheney was Darth Vader in the White House, ruthlessly seeking power and using American military forces to upend the sovereignty of foreign nations, that was fine with Bush.

After one debate prep at the vice presidential residence during the 2004 campaign, Cheney's chief of staff Scooter Libby and Patel took Cheney aside. "Sir, have you looked at your numbers?" Libby asked him. His approval rating had fallen dangerously low as he became the face of the Iraq War and the botched prewar intelligence about Iraq's chemical and nuclear weapons program used to justify it. (On March 16, 2003, four days before the U.S.-led invasion of Iraq, Cheney famously told *Meet the Press* host Tim Russert, "I really do believe that we will be greeted as liberators . . . The read we get on the people of Iraq is there is no question but what they want [is to] get rid of Saddam Hussein, and they will welcome as liberators the United States when we come to do that.")

Cheney was not interested in poll numbers. "No, not really," he replied.

"You might want to take a look," said Libby, "they're pretty bad."

Cheney asked for the numbers. "Should we get out and do more press with you?" Libby asked.

Cheney laughed and replied, "You've got to remember, no one is voting for vice president. I really don't care."

And he did not particularly care how he was viewed in Congress, especially among Democrats. During the Senate class photo in 2004 Senator Patrick Leahy, a Democrat from Vermont, and Cheney got into a heated exchange on the Senate floor, where swearing is against the rules. (The Senate was technically not in session at the time.) Cheney, as president of the Senate, was there for picture day and scolded Leahy over his criticism of Cheney's ties with Halliburton, one of the largest oil extraction companies in the world. Democrats at the time were suggesting that he helped Halliburton—where he was CEO from 1995 to 2000—win lucrative contracts to rebuild Iraq. "Fuck off," Cheney told Leahy.

Cheney is just five years older than Bush, but he had the kind of Washington gravitas and influence that Bush was sorely lacking. Cheney's assistant Lucy Tutwiler recalls sitting at her desk outside of Cheney's West Wing office one day when the vice president was traveling. Her desk faced the hallway and she watched in surprise and amusement as Bush rode a bike down the hall in her direction. He stopped at her desk and asked, "Where's the vice president?" He wanted to show off the bike, which she said was probably a gift, to Cheney. When she told him Cheney was out, Bush took a lap down to the Oval.

Richard Bruce Cheney was born in 1941 in Lincoln, Nebraska, and has a long history in government rivaling George H. W. Bush's. He served as President Gerald Ford's chief of staff and President George H. W. Bush's defense secretary, with six terms as a Wyoming congressman in between, including a stint as House minority whip. In December 2008, after Barack Obama won the election, Cheney attended a private breakfast with incoming White House chief of staff Rahm Emanuel. Cheney had been invited by then–White House chief of staff Josh Bolten, who gathered together thirteen of the living sixteen men who had served as chiefs of staff. Emanuel went around the room seeking advice, and Cheney told him, with a mischievous grin, "Whatever you do, make sure you've got the vice president under control." The room exploded into laughter. (Trump's former chief of staff Reince Priebus also reached out to Dick Cheney for advice.)

Cheney firmly believes in the example Richard Nixon set on the campaign trail when he was Eisenhower's vice president. "I found the best way to help was always to campaign for the President and never for myself," Nixon wrote to Bob Dole, when Dole was Gerald Ford's running mate. "Also remember that you should always attack *up* and never *horizontally.*" That was precisely the strategy Bush and Cheney used in the 2004 campaign when Cheney was sent out to hammer Democratic challenger John Kerry—an assignment he thoroughly enjoyed. The Bush campaign tried to create a Cheney-Kerry debate so that Bush could hover above the fray. "On a daily basis we would tweak Kerry and Kerry would always reply because Cheney has this gravitas," said Patel. "We thought it was hilarious. It's hard to ignore it if Cheney is coming after you."

A week after Kerry and his running mate, John Edwards, were nominated in 2004, Cheney gleefully mocked Kerry (not Edwards) at a campaign event in Dayton, Ohio. Cheney used Kerry's own words against him and mocked him for saying he would wage a "more sensitive" war on terror as the crowd roared with laughter. Great presidents like Abraham Lincoln and Franklin Roosevelt "did not wage sensitive warfare," Cheney said. "Those who threaten us and kill innocents around the world do not need to be treated more sensitively. They need to be destroyed."

Cheney's vice presidency was like none before. He received the top-secret President's Daily Brief, or PDB, a ten-to fifteen-page intelligence summary compiled by the CIA, in the early morning hours at the Observatory, typically before the president himself saw it. The PDB has been referred to as a newspaper with the smallest circulation in the world. His chief of staff, Scooter Libby, was usually with him when he got the briefing. Vice presidents, Cheney argues, should be playing an active role in national security issues, if for no other reason than to prepare them to step into the presidency in case of an emergency.

"I can see other vice presidents who didn't have that background or that interest. My personal view is that they need to develop that, that that's a key part of being president." Cheney sometimes even reshaped the PDB, telling intelligence briefers to present certain material up front to the president. When he received the book, there would be a tab, he said, and "behind the tab" would be things he had specifically asked about. Cheney said he had more time to study the briefing book than Bush did, given that the president was often preoccupied with other official, and sometimes mundane, responsibilities. According to his successor, Joe Biden, Cheney had a steel door put on a room in the Observatory that he used as a sort of vice presidential Situation Room so that he could participate in classified meetings at any time. He had an insatiable appetite for work and was up at 4:30 a.m. reading, and received a private intelligence briefing between 6:30 and 7:00 a.m. An hour later, he would be in the Oval Office for the president's daily briefing. When he was traveling, Cheney had with him a solar-powered battery station that linked him by satellite secure line to the Situation Room so he could go over the PDB. No matter where he was, he had a CIA briefer with him, including when he went on fishing trips at the remote south fork of the Snake River along the Wyoming-Idaho border. President Trump, it turns out, rarely reads the PDB and instead relies on the oral briefings.

Much of Cheney's influence as vice president came from his stewardship in the administration's earliest days after 9/11. As a young, ambitious Republican congressman, Cheney had taken part in the Reagan administration's clandestine exercises to prepare for a global nuclear war with the Soviet Union. At least once a year Cheney and his friend Donald Rumsfeld, who was a business executive at the time, would be ferried away to an undisclosed location where they helped plan for a worst-case scenario where the president, vice president, and Speaker of the House were dead or incapacitated. The annual doomsday operations helped prepare Cheney for his role as a wartime vice president. After the second plane hit the World Trade Center, one of Cheney's Secret Service agents grabbed him by his belt and moved him to the Presidential Emergency Operations Center (PEOC) in the basement of the White House, where other senior aides were gathering. Another plane, Flight 77, was believed to have been hijacked and was heading toward Crown, the code word for the White House. In the basement Cheney recalled lockers being opened and weapons being passed out among the agents. Once in the basement tunnel, Cheney called Bush, who was on Air Force One after leaving an event at a Florida school, and advised him not to return to the White House. "He didn't like that at all," Cheney said. "In the end he agreed to do it."

National Security Adviser Condoleezza Rice; her deputy, Stephen J. Hadley; economic adviser Lawrence B. Lindsey; White House aide Mary Matalin; Cheney's chief of staff Scooter Libby; and the vice president's wife, Lynne, were gathered around the television watching in horror. When the first tower fell there was a collective groan in the room, but Cheney was quiet, his eyes closed, saying nothing. Eventually the Secret Service asked several staff members to leave the PEOC because they were putting a strain on the air-conditioning and the piped-in oxygen supply in the relatively small space.

That day came to define Cheney's relationship with Bush. The night of September 11, the Cheneys were evacuated to the secluded Camp David, where they watched replays of the day's horrific events in the presidential cabin, Aspen Lodge. It was while at Camp David that Cheney made notes to relentlessly go after the terrorists and those who support terrorism. His remarks on NBC's *Meet the Press* on Sunday, September 16, 2001, began to define what would become the "war on terror." "We have to work the dark side, if you will," he said. "Spend time in the shadows of the intelligence world. A lot of what needs to be done here will have to be done quietly." Bush and Cheney even sat down together for more than three hours with members of the independent commission investigating the 9/11 attacks. Bush said it was important to him that they appear together so that members of the commission could "see our body language . . . how we work together."

"It was the president's suggestion," Cheney said. "I thought it was a good one. That our actions together on that day were as part of a team."

———

When the chairmen and ranking members of the House and Senate Intelligence Committees arrived at the White House for a briefing on surveillance shortly after September 11, they were surprised to learn that they would be meeting with Cheney and not the president. The president pointedly told then-senator Bob Graham of Florida, who chaired the intelligence committee, that the vice president would be the main point of contact on intelligence matters.

"Prior to 9/11 we treated terrorism as a law enforcement problem," Cheney said, pointing to the 1993 World Trade Center bombing and the conviction in federal court of the four men responsible for the bombing that killed six people. "If you bought my view of the world, 9/11 justified a far more robust response and that it was an act of war. If you compare it to Pearl Harbor it was worse than Pearl Harbor, worse in a sense that we lost 3,000 people compared to 2,400 people. Pearl Harbor was a territory of the United States, this had been a strike at the homeland," Cheney reflected. "This was an act of war and that meant in my mind tough aggressive measures designed to go after those who were responsible and to operate more aggressively than we ever have before."

Cheney installed his top staff as members of the president's staff as well: Mary Matalin was both his communications director and an assistant to the president; Scooter Libby was an adviser to both Cheney and to the president. Cheney's approach to the vice presidency was in part shaped by his training as Ford's chief of staff. He rarely spoke during meetings and often was disconcertingly quiet, offering other aides and Cabinet members little sense of where he actually stood on an issue. This was done entirely on purpose. When Cheney was alone with the president, he had tremendous influence. "As chief of staff I felt the need to be an honest broker, not to be the guy advocating cutting the defense budget publicly, because if I'm doing that then the secretary of defense is going to find some way to get around me to get to the president and then

During the 1992 campaign, the Clintons and the Gores were like baby boomers on a double date. But the similarities between Al Gore and Hillary Clinton eventually caused strain inside the White House.

Bill Clinton's affair with White House intern Monica Lewinsky severed a bond between the two men. "For almost all those eight years the relationship was one between brothers," Gore said. Hillary and Gore were horrified by the betrayal, but it stuck with Gore's wife, Tipper, most.

Hillary Clinton ran on the Clinton-Gore record and won her New York Senate race in 2000 by double digits. As president of the Senate, Gore had to swear in Hillary after his own devastating defeat.

Gore and Clinton talk in the Oval Office days after his 2000 loss. Not long after the shocking results of the 2016 presidential election, Gore called Hillary to commiserate.

On 9/11, when the first tower fell, Dick Cheney was watching television with other top Bush advisers. There was a collective groan in the room but Cheney was quiet, closed his eyes, and said nothing.

Shortly after 9/11, Lynne Cheney gathered her small staff on the veranda at the vice president's residence. She brought a copy of David Brinkley's *Washington Goes to War*, a book about the capital at the beginning of World War II. Here the Cheneys are on board Marine Two on 9/11.

In George W. Bush's second term, Cheney's power waned considerably. One Bush aide said, "He was the grouchy uncle yelling at everyone to get off the grass."

Bush and Cheney in the Residence elevator, September 11, 2008.

Dick Cheney on Joe Biden: "Personally we always got along fine.... I don't see him as a heavyweight."

Joe Biden on Dick Cheney: "Give me back the neocons. Cheney was really smart and really informed. I think he was wrong, but not because he wasn't informed."

On election night in 2008, Biden's mother, Jean, then ninety-one years old, reached for Obama's hand as they walked on stage in Chicago's Grant Park. "She wasn't supposed to walk out," Biden recalled. "And she says, 'Come on, honey, it's going to be OK,' to Barack."

Inside the White House, Obama was nicknamed "Mr. Spock," after the human-alien philosopher on *Star Trek*. Biden helped Obama connect with his emotions. Here they talk before Obama's first inauguration, January 20, 2009.

"That 'bromance' thing is real man," Biden says. Here Biden and Obama attend a rally in New Hampshire during the 2012 campaign.

Obama's top advisers gather to watch the Navy SEAL raid that killed Osama bin Laden. Though he said he advised Obama against the raid in a meeting with other aides, Biden insists that he privately told him to trust his instincts.

Unlike every other modern presidential example, the relationship between Biden and Obama grew stronger over their eight years together.

ABOVE LEFT: President Barack Obama laughs with Vice President Joe Biden in the Oval Office in 2015. ABOVE RIGHT: Biden, with Julia Louis-Dreyfus, star of the HBO show *Veep*, in Biden's West Wing office.

Obama looks over Biden's remarks announcing his decision not to run for president. Biden was still grieving after the death of his son Beau. "He's the only person outside my family I ever told about Beau," Biden said of Obama, his eyes filling with tears.

Donald Trump announces Mike Pence as his running mate on July 16, 2016. Trump's daughter Ivanka and her husband, Jared Kushner, look on. "We've been praying for you," the Pences told Donald and his wife, Melania, shortly before the announcement.

Biden has served as a conduit between Pence and foreign leaders who he says are "diffident" about Trump.

Karen Pence seems like an unassuming midwestern mother of three, but she has been wildly underestimated.

Pence is deeply religious and hosts a lunchtime Bible study, led by an evangelical pastor, at the Eisenhower Executive Office Building.

Joe Biden, Walter Mondale, and Mike Pence in the vice president's Senate office just off the Senate floor on January 3, 2018. Pence has sought advice from several former vice presidents.

Plotting his future: Vice President Pence is playing the long game. When he goes to Capitol Hill to "touch gloves," as he likes to say, he insists on walking through the Rotunda so that tourists can get photos with him. For now, he is Trump's loyal soldier. "His job is defined by one person," said Pence's first vice presidential chief of staff, Josh Pitcock, "and the most important thing is to protect that relationship."

you lose control of the process." He had spent decades studying the West Wing. Cheney jokes about being the oldest guy in the West Wing and the only person the president could not fire.

Cheney attended weekly Senate Republican policy lunches on the Hill but was as inscrutable in those as he was in West Wing meetings. The Republican leader would open the lunch up and ask the vice president to speak. "Sometimes I did, sometimes I didn't," Cheney said, with a Cheshire-cat-like grin. "Cheney was very disciplined about the nature of his relationship with the president and how he interacted with the president in front of others," said Cheney aide Patel. "He wouldn't weigh in at meetings very often, he didn't want people to know if there was daylight between him and the president on issues. Even with his staff he didn't come back to us with a big readout on how the president reacted to something in a private conversation. Cheney thought that would have been improper." He withheld his personal opinion often until he was alone with the president, and the president was free to take it or not.

Cheney developed and pushed for the policy of trying foreign terrorism suspects in closed military tribunals, thereby stripping them of access to civilian courts. Attorney General John Ashcroft, national security adviser Condoleezza Rice, and Secretary of State Colin Powell all disagreed, but Cheney used his stature and his relationship with the president to circumvent them all. "In the aftermath especially of 9/11, we needed to get things done, and on occasion I would use the position I had, and the relationship with the president I had, to short-circuit the system. No question about it," says Cheney. As Ford's chief of staff he knew the system well and knew exactly how to maneuver around it. Ashcroft was especially angered by the fact that the Justice Department would have no role in choosing which alleged terrorists would be tried in military commissions. When Ashcroft went to the White House to voice his concerns, he could not get an audience with the president and instead found Cheney seated in the Roosevelt Room. As the top law enforcement officer in the country he was incensed and told Cheney and others in the room that he would need a role in the tribunal process. On November 14, 2001, the day after Bush signed the order creating the military tribunals, Cheney told the U.S. Chamber of Commerce that terrorists do not "deserve to be treated as prisoners of war." The president did not yet support that position, but after more than two months he decided in Cheney's favor: the Geneva Conventions should not be applied to Taliban and al-Qaeda fighters who were captured in battle. (The Geneva Conventions protect civilians and soldiers in war zones and have been in place since 1949.)

Halfway through preparations for the reelection campaign in 2004, Cheney recalls making the president an unusual offer. "Look, one of the things you can do to try and change the landscape out there, maybe gain a few points, is to get yourself a new running mate," Cheney said. "I want you to know I'm prepared to step down in a minute if you decide you want to do that—it's fine with me, love the job and am happy to proceed but you really need to have the option." Cheney said he approached Bush three times and the first two times Bush "just sort of blew it off," but the third time he went away and thought about it for a couple of days. Bush came back and told him, "You're my guy, Dick." Cheney still wasn't sure. "I watched the situation in '92 with his dad [and his decision to keep his vice president, Dan Quayle, on the ticket during his reelection campaign] where I think there was merit to the argument that if he had been willing to change, not for me necessarily, but if he had gotten new talent that might have helped his dad."

There were occasions when Cheney found it "convenient" to use his constitutional authority and be a member of both the executive and legislative branches. He tested the limits of vice presidential power when he intentionally contradicted the president in 2008, at the tail end of their tenure. Cheney expressed his displeasure with a Justice Department brief that he thought was not strong enough in protecting the Second Amendment right to bear arms in a case before the Supreme Court, *District of Columbia v.*

Heller. Texas senator Kay Bailey Hutchison asked Cheney whether he wanted to add his name to a pro-gun friend-of-the-court brief signed by 250 pro-gun House members and 55 senators who were dissatisfied with the administration's position. He asked his counsel, David Addington, who told him that he was legally allowed to do so. Without telling anyone in the West Wing, Cheney signed on, using his position as president of the Senate to take on a purely legislative role. In a 5–4 ruling, the Supreme Court struck down the District of Columbia's thirty-two-year-old ban on handguns. The West Wing was caught entirely off guard when Hutchison released the brief bearing the vice president's name.

"The immediate reaction downtown—the president never said a word about it, I think he agreed with me substantively because we campaigned on that—but Josh Bolten, the chief of staff, wanted to call a process foul . . . They had it in their minds I'm part of the administration et cetera and they hadn't really studied beyond that," Cheney said, more amused than angry. Bolten told Cheney he planned to talk with Addington. Addington was not intimidated, and when Bolten made clear his displeasure with the vice president coming out against the president's position, Addington reminded him that he worked for the vice president and that his paycheck came from the Senate, not the White House.

"Understood," Bolten said, "but if we have another episode like this, I will make sure that all of your belongings and your mail are forwarded to your tiny office in the Senate and you won't be welcome back inside the gates of the White House."

Addington and his boss were not worried. "Dave made it clear to them that I had two hats," Cheney said, "half my salary and budget came from the Hill and half from the executive branch."

————

In the late afternoon of Saturday, February 11, 2006, Cheney, an experienced hunter, used a 28-gauge Italian-made shotgun to shoot at a covey of quail at the Armstrong Ranch in south Texas.

He shot some two hundred lead birdshot pellets before he noticed someone in a bright orange vest with only his torso visible standing in a gully about thirty yards away. It was seventy-eight-year-old Harry Whittington, a prominent Austin lawyer and campaign contributor who was part of the small party hunting quail that weekend. Some of Cheney's pellets hit Whittington in the chest and face. Cheney ran to Whittington's side as he lay bleeding but still conscious on the ground. Before he passed out, Whittington remembered smelling gunpowder. Cheney always travels with a medical team, and they administered first aid. "I said, 'Harry, I had no idea you were there,'" Cheney recalled. "He didn't respond."

Patel was the most senior Cheney aide on the trip. He was at a nearby hotel when he got the call from the vice president's photographer, David Bohrer. "We have a problem," Bohrer told him. "The boss shot somebody." A piece of birdshot lodged near Whittington's heart muscle and triggered a mild heart attack in the hospital. Whittington did not look good when Cheney went to visit him there. "We were trying to track his kids down so they didn't think their dad was dying and the White House guys were just pissed that we weren't getting in front of the story press-wise," Patel recalled. "Our view was that this was a hunting accident, this was not a political story, we're not going to deal with the White House press corps, it's a local hunting accident."

News of the shooting did not emerge until Sunday when the *Corpus Christi Caller-Times* reported the incident and said that Whittington had broken away from the group of hunters without announcing himself. Blaming the victim caused outrage, and the whole incident became comic fodder for late-night comedians. When Whittington was released from the hospital, he actually apologized to Cheney. "My family and I are deeply sorry for all that Vice President Cheney and his family have had to go through this past week," he said. Whittington, who still has pellets in his neck and right cheek, did not pick up a shotgun again for a decade afterward.

But because of Cheney's dangerous obsession with secrecy—he declined to disclose the names or the size of his staff and usually had no public calendar—the White House spent days grappling with questions about the accident and why it was not revealed immediately. The only information provided was that there had been a hunting accident and that Cheney was fine. Cheney's aides left out the key detail that it was Cheney who was the person responsible. First Lady Laura Bush, who rarely weighed in on politics and was in Italy for the Winter Olympics, was angry that the White House was not getting the news out sooner, telling her chief of staff Anita McBride to call her husband's chief of staff Andy Card and tell him to get the story out as soon as possible. White House press secretary Scott McClellan and Bush counselor Dan Bartlett kept calling Patel Sunday morning but, remarkably, their calls were not immediately returned. Patel said the decision to call the local paper was a reflection of Cheney's disdain for the White House press corps, and he described West Wing staff as being "enraged" by Cheney's decision to break the story to a local paper.

Bartlett finally got on the phone with Cheney that Sunday, February 12, and told him they needed to gather a so-called pool of reporters in Texas so that it would not look like they were trying to bury the story by giving it to a small newspaper. Cheney was silent, then said, "This is how we're going to handle it." Finally, on February 15, four days after the accident, Cheney did an interview with Fox News's Brit Hume. In it, he was uncharacteristically emotional and said, "It's not Harry's fault. You can't blame anybody else. I'm the guy who pulled the trigger and shot my friend." He added, "The image of him falling is something I will never be able to get out of my mind. I fired, and there's Harry falling. And it was, I'd have to say, one of the worst days of my life, at that moment."

In retrospect, Cheney said, it was not the smartest way to handle the situation, but there were not many options: "There's no good way to handle a sitting vice president shooting someone," Patel reflected. "And there's no way to make it not be a giant, national story."

———

The embarrassing hunting episode was the beginning of the end of Cheney's outsize influence on Bush. He was starting to feel like some of the less powerful vice presidents who preceded him. Bush adviser Karl Rove used to call Cheney the "Management"—as in, "Better check with Management"—but by the second term Cheney was no longer the font of wisdom he had been in the beginning, in the days following 9/11, when Bush relied on his foreign policy and government experience. He admits his waning influence himself. "By the time you get to the second term he's [Bush is] an expert," Cheney said. "I think he was more assertive himself and had more confidence in the second term, and I don't find that surprising."

The hunting accident was humiliating for the vice president, even though he made light of it, telling an aide, "Now *that* was a good one!" when he noticed a graduate wearing a bull's-eye on his mortarboard when Cheney gave the commencement address at Louisiana State University in May, less than four months after the shooting. Less than six months later, Cheney gathered with Treasury Department officials, Bush's counsel Harriet Miers, Rove, and Press Secretary Tony Snow in Chief of Staff Josh Bolten's office to discuss how to handle a story the *New York Times* was about to publish on a secret U.S. program to monitor terrorist financing shortly after September 11. Reporters at the *Wall Street Journal* and the *Los Angeles Times* were chasing the story but were behind the *New York Times* on their reporting. Bush administration officials were deciding whether to work with the *Times* or to work with a "friendly" reporter who would not sensationalize the story, as they thought the *Times* would. When this was suggested, Cheney, who as usual was sitting quietly and watching, chimed in with perfect timing and a half-smile, "You mean like the *Corpus Christi Caller-Times.*"

The room erupted in laughter as even Cheney acknowledged that he had made a mistake.

During his first week on the job in Bush's second term, deputy press secretary Tony Fratto recalled working on a statement that, like all statements, had to be sent to other departments for review and editing before it could be emailed widely. He was new to the process and was surprised when two changes came back rewriting a sentence in the second paragraph—one from Rove and one from the vice president's office. Both sentences were rewritten in entirely different ways. Fratto called senior Bush aide Dan Bartlett and asked, "Who big foots who?" Bartlett told him, "*You* big foot." Fratto did his best to reconcile the two but remembers how striking it was that Bartlett did not tell him to come down on either side, a small indicator of Cheney's waning influence.

During the last couple of years of the Bush administration, Cheney seemed drained, in part because of his heart issues. A nod of the head at a meeting or a rare pointed question—all things aides used to look for to indicate where he stood on an issue—became almost irrelevant. Now Bush's aides were occasionally mocking him. After one meeting in which Cheney dozed off, Bush himself laughed. "Did you see?" Bartlett asked Bush. "Yes," Bush replied. "I couldn't look at him." Ambassador Christopher Hill recalled another meeting when Cheney fell asleep in the Oval Office. Secretary of State Condoleezza Rice took Hill aside afterward and said, "Did you notice that? That happens a lot these days." She was speaking about Cheney's health but she was also speaking metaphorically—he did not matter as much as he used to.

During Bush's second term Rice replaced Cheney as Bush's closest adviser, sidelining the vice president. On June 17, 2007, Bush gathered his national security staff for an evening meeting in the second-floor Yellow Oval Room in the White House. The meeting was to discuss the al-Kibar nuclear reactor in Syria; since the end of 2006 Israel had been telling U.S. officials about possible nuclear activity in Syria. There was evidence from intelligence agencies that

North Korea was helping the Syrians build the reactor. Israeli intelligence operatives had managed to get photographs taken inside the top-secret plutonium reactor, and of North Korean workers at the site. The reactor's only purpose, Israel concluded, was to build an atomic bomb.

Deputy national security adviser Elliot Abrams argued for Israel to bomb the plant. Cheney was alone in his ardent support for the United States to bomb the plant. He insisted that Bush had imposed a red line on North Korea in October 2006 when it tested an atomic bomb, and this was clearly a case of nuclear proliferation. No one raised their hand when Bush canvassed the room and asked, "Does anyone here agree with the vice president?" Cheney recalled, "I was a voice in the wilderness at that point." He had long urged the intelligence community to look into a link between North Korea and Syria, and here was evidence that he was correct. "They [Bush and his advisers] were snake bit because of the war in Iraq," Cheney said. "I really believe the history books would show that would have been the right call." At a later meeting, Secretary of Defense Bob Gates said, half in jest, "Every administration gets one preemptive war against a Muslim country, and this administration has already done one."

Christopher Hill, who was then head of the U.S. delegation to the six-party talks aimed at peacefully resolving North Korea's nuclear weapons program, says that Cheney was not influential on North Korea during the second term. "Cheney was a critic and obviously very negative in his approach to the issues that were essentially being run by Condi Rice and the president. It was a little odd having a vice president being so critical. He was very negative about everything without any real role with it." Hill and other Cheney detractors argue that Cheney's fatal flaw inside the White House was his condescension toward Bush. He viewed Bush as a politician, whereas he had far more foreign policy experience. But in the second term Bush did not seek his advice nearly as often.

Hill, who would be named U.S. ambassador to Iraq during President Obama's first term, said the body language between Bush and Cheney was far different from that between Obama and Biden. "Bush would say things, and I'd look over at Cheney and he just had his head down. He didn't seem impressed. There was no effort to reinforce the president's point," Hill recalled. "I never saw him reinforce a point that the president had made. I felt it was 'I know best, I don't need to reinforce *his* point, he needs to reinforce *my* point,' whereas Biden, I think, was just far more overtly respectful of his boss." But Cheney's penchant for being quiet during meetings does not mean he did not respect Bush or the office he held. Bush called Cheney "Dick" while Cheney always called Bush "Mr. President."

Part of Cheney's waning influence was the absence of his longtime friend and mentor Donald Rumsfeld, the man Bush personally asked Cheney to fire as secretary of defense in 2006 (most presidents avoid doing the actual firing themselves). One aide said, "Rumsfeld and Cheney together were one plus one equals three." They were a powerful duo, and without Rumsfeld, Cheney truly was "a voice in the wilderness."

Hill says that former secretary of state Henry Kissinger told him that after meeting with Bush, Kissinger walked out, saw Cheney, and without any appointment sat down with Cheney and talked for an hour. Kissinger's take on this, Hill said, was that it was shocking to sit down with a vice president with no appointment, no notice, and to have an hour with him. It suggested that maybe Cheney wasn't very busy during the second term. It stood in stark contrast to the early days of the administration, like in 2003 when Bush asked to be alone with Cheney before launching the first strike that started the Iraq War. "He was marginalized so much [during Bush's second term] that it didn't really matter," said Hill. "He was the grouchy uncle yelling at everyone to get off the grass."

Cheney's chief of staff I. Lewis "Scooter" Libby spent years over-seeing a mini fiefdom in the West Wing with Cheney as its leader. Libby had an unprecedented triple title: Cheney's chief of staff and national security adviser as well as assistant to the president. In 2007, he was found guilty of lying about his role in the leak of undercover CIA operative Valerie Plame's name to the press and was convicted of perjury and obstruction of justice. Though Bush commuted the sentence, which meant Libby did not have to serve thirty months in prison, Cheney insisted then, and he still does to this day, that Libby should have been granted a full presidential pardon. Friends say they do not think anything bothered Cheney more than what he described as leaving a man on the battlefield.

"I did everything I could [to convince Bush to pardon him]. Without question it was the most tense element of the relationship between myself and the president," Cheney said. "I still think he made a major mistake when he didn't pardon Scooter . . . I still want to see a pardon, that's the only thing that I think is justified by the circumstances . . . It's a travesty . . ." Cheney continued, "We had a very, very heated exchange the last time we had lunch in the White House the week of the inaugural [Barack Obama's 2009 inauguration]. It's a subject we don't discuss." But the deci-sion to not pardon Libby clearly weighed on Bush. Riding together in the motorcade to the Capitol with his successor, Barack Obama, Bush advised him to come up with a pardon policy early on in his administration and to stick to it, no matter what. After Obama's inauguration, Cheney and Libby, his longtime right-hand man, did not see each other for more than a year.

On Inauguration Day 2009, Cheney was in a wheelchair because he said he had thrown his back out packing boxes at the Observa-tory. Cheney joked to Biden, his successor, "Joe, this is how you're liable to look when your term is up." The relationship between Cheney and Bush did what those of most of their predecessors' have done: it deteriorated over time. They had been through hell together and they respected one another, but friends say they do not see each other often and they are not close. Since leaving the White House, Bush has described their relationship as "cordial." "But he lives in Washington and we live in Dallas."

Their partnership was never a warm friendship—it was a business relationship that was strengthened by their unity in the aftermath of 9/11 and it soured by the end. Since leaving office, Bush has become livid at the suggestion that Cheney had an outsize role in his White House. In a recent book, Bush said that Cheney and Secretary of Defense Donald Rumsfeld "didn't make one f–ing decision" while he was president. "The fact that there was any doubt in anyone's mind about who the president was, blows my mind," he said. But there was a time not that long ago when Bush would have given his vice president more credit. In 2004, at a stop in Raleigh, North Carolina, after Kerry announced John Edwards as his running mate, a reporter asked Bush what he thought of him. "He's being described today as charming, engaging, a nimble campaigner, a populist and even sexy," the reporter said. "How does he stack up against Dick Cheney?"

"Dick Cheney can be president," Bush answered, sounding annoyed by the question. The answer seemed so obvious. "Next," he said, pointing to another reporter.

XII

Fanboy:
The Love Story of Joe and Barack

He was kind of a fanboy, honestly.

—A FORMER TOP BIDEN AIDE ON VICE PRESIDENT JOE BIDEN'S
RELATIONSHIP WITH PRESIDENT BARACK OBAMA

*That's the job of the vice president, you're supposed to throw yourself in front
of the train. That's one of the reasons I didn't want to do it in the beginning.*

—JOE BIDEN ON HIS RECOMMENDATION AGAINST SENDING
IN NAVY SEALS TO KILL OSAMA BIN LADEN

He is fearless about emotion . . . He understands real pain.

—VALERIE JARRETT, BARACK OBAMA'S SENIOR WHITE HOUSE ADVISER
AND CLOSE FRIEND OF BARACK AND MICHELLE OBAMA'S

Joe Biden was up very late the night before he made his October 21, 2015, Rose Garden announcement that he would not be seeking the presidency. He was on the phone with a close friend, South Carolina state representative James Smith—South Carolina would be a linchpin for Biden if he decided to run. Biden and Smith would speak on the phone four times in forty-eight hours before the decision was made, ending months of intense media speculation. Starting at about 11:00 p.m., Smith said, he talked with Biden for thirty minutes, and Biden read him his speech in the emotional call from the vice president's residence. Smith was in Fort Worth, Texas, sitting on the curb in front of his hotel. "There was all this pent-up—I lost my cool, I said, 'Dammit, Joe Biden, you have to run because no one else is saying these things.'" Smith was convinced that Biden was the one person who could bring the country together. Biden told him that he was going to sleep on it. "He had not made up his mind as of 11:30 p.m. the night before," Smith recalled.

Biden and his close aides, Mike Donilon and Steve Ricchetti, had been far enough along in their plans for his presidential run that they were thinking about possible running mates. In late August, Biden invited Massachusetts senator Elizabeth Warren for an unannounced Saturday lunch at the Naval Observatory, an indication that she was being seriously considered as a running mate. Biden wanted to run a Bernie Sanders–type campaign, aides say, with free college tuition being one of its major appeals to working-class voters. But, in the end, Biden decided he could not do it—his family needed him after the death of his eldest son, Beau, from brain cancer less than six months earlier. Beau was just forty-six years old and a talented politician who Biden joked had more in common with Obama's calm temperament on the political stage than his own. When Biden's speechwriters needed him to keep a speech short they would tell him to imagine the disciplined Beau standing at the teleprompter. More than anything Biden wanted Beau to be proud of him.

When Beau did television interviews as Delaware's attorney general, Biden's staff always made sure to send him a transcript afterward. He would stop whatever he was working on, read what his son had said on his smartphone, and summon his nearest aide. "Look how good this kid is," he would say, showing them his phone. Beau's loss was deeply felt by the Obamas, too. The day after he passed away, the Obamas canceled a reception for Ford's Theatre so they could visit the Bidens at the Naval Observatory. Biden saw Beau as his heir apparent. Biden's former chief speechwriter Dylan Loewe said, "If Joe Biden runs in 2020, in some ways it would be because his son couldn't."

Beau's death came after years of illness and worry. In May 2010 Beau, then forty-one, had a stroke that is thought to have been a precursor to the brain cancer that would take his life. Obama's senior strategist David Axelrod recalled the day of Beau's stroke. During private meetings with his staff, Axelrod said, Obama was almost always fully prepared and engaged in the discussion. On that particular day, after hearing the news, the president stared out the window, clearly distracted, as his advisers discussed policy. Finally, Obama said quietly, "I don't know how Joe's going to go on if something happens to Beau." Later, when Biden was back in his West Wing office, Axelrod saw Obama sprint down the corridor. "I poked my head out and I saw the two of them embrace outside the vice president's office." Obama, in shirtsleeves, grabbed Biden by the shoulders, and asked him, "Joe, Joe, is he OK? Is he OK?" Axelrod said, "I think that says everything about their relationship." Three years later Beau was admitted to MD Anderson Cancer Center in Houston and was diagnosed with brain cancer. Obama told a key aide: *Handle this.* It was as if he was talking about his own brother; he wanted to make sure Beau and his family were given as much privacy as possible and had the best care possible.

Beau's death on May 30, 2015, was the dominant factor in Biden's decision not to run. When Beau was near the end of his life, the Bidens wore blue bracelets—blue was Beau's favorite color—with "WWBD" engraved on them—What Would Beau Do. Biden was distracted by his grief. He did not tell most of his staff that Beau's diagnosis was terminal, but they noticed that as Beau was being treated for cancer Biden lost weight and often looked exhausted. Before he passed away, Beau made one request of his father: "You've got to promise me, Dad, that no matter what happens, you're going to be all right." Biden began wearing Beau's rosary around his wrist after he died, as a way of reminding himself of that promise.

After the music was selected for Beau's funeral, Biden went alone into his office, shut the door, and played the songs on full volume over and over again so that he could numb himself and not break down crying in the church. At the funeral, Biden stood next to his son's casket at the front of St. Anthony of Padua Church in Wilmington. When one staffer hugged Biden and told him how much he loved Beau, the vice president replied, "Beau loved you too, buddy." It seemed to the staffer like Biden was on autopilot and was just getting through what he thought he owed his son. Another former Biden aide compared Biden to the godfather of a large Irish Catholic family holding everyone together. He mourned openly, and that public expression helped him manage his grief.

Two months after Beau's death, an aide carefully orchestrated a call from Biden to *New York Times* columnist Maureen Dowd. In the resulting op-ed, Dowd described Beau, near the end of his life and with his face partially paralyzed, begging his father to run and "arguing that the White House should not revert to the Clintons and that the country would be better off with Biden values." But Biden's friends, an aide said, "were trying to get the guy up in the morning so he could keep the family together, so he didn't put a bullet in his head." Still, it was important to Biden that people saw him as someone who could have been president and that, had he chosen to run, he just might have won. The fact that so many people were rooting for him at the worst moment in his life, when he needed it the most, provided some small relief. Even former Republican senator Bob Dole took Biden aside and said, "Joe, you better get in this race."

One aide described Biden's philosophy: "In politics and life you're either on your way up or on your way down." And he never wanted to be on his way down. By putting his name out there, he was showing he was still powerful and that reporters would still write about him. "If you've been a politician for forty years," a former Biden aide said, "that's part of your ego, and you measure yourself in part by the number of inches you get in the newspaper every day." Biden's family members were furious that he was getting slighted in the conversation about who should run for president in 2016, chief among them Beau. Members of his close-knit and loyal staff were frustrated, too. "I thought the remedy for Obama was twenty steps down the hall," said one longtime Biden adviser. It was Biden, they believed, who could best reach out to members of the white working and middle classes, not Hillary Clinton. Even top Democratic Party officials were frustrated by the Clintons' hold on the party. At one point during the campaign, when Clinton had a fainting spell brought on by pneumonia, then–Democratic National Committee chair Donna Brazile considered initiating a process that would have replaced Clinton and her running mate, Tim Kaine, with Biden and New Jersey senator Cory Booker.

Biden's sons, Beau and Hunter, were his strongest and most loyal champions. They always stuck by their father's side—in 1988, when Biden decided to drop out of the presidential race, Beau urged him not to give up. "Dad, if you leave, you'll never be the same."

In 2016 President Obama made the unusual decision not to endorse Biden, his own vice president, for the presidency. The deck was stacked against Biden with top Democratic donors already firmly in Clinton's corner. Still, senior advisers to Obama now say that Biden could have defeated Donald Trump if he had gotten past Clinton in the Democratic primaries. But Clinton's nomination was a fait accompli that even the president could not stop. Obama thought Biden had been through too much personally and needed to give up on running, but he did not directly make that case to him. Instead he cautioned Biden to remember the rigors of the campaign and the deep grief that he and his family were going through. Privately, Biden thanked White House press secretary Josh Earnest for not directly answering reporters when they asked, almost daily at one point, whether he would be a candidate in 2016. He was grateful for the breathing room to make the decision.

Obama saw the significance of passing on the presidency from the first black president to the first female president. Former White House deputy chief of staff Alyssa Mastromonaco said Obama never lost sight of one important fact: "He knew he was surrounded by people whose dream he was living, with Hillary Clinton on one side of him and Joe Biden on the other." That thought, she said, "never escaped him." Everyone inside the White House assumed Biden would run. "He was our guy. If he was going to do it we were going to be there for him," said Mastromonaco. "There was a deep abiding Biden loyalty." Biden thought he should have been consulted before Hillary ran. After all, they had known each other for years—Biden admired Bill Clinton in particular and recognized him as the only national politician better than he was at connecting with people—and he had encouraged Obama to name her secretary of state. Biden viewed Obama on a higher plane that was out of his league, but Hillary and Bill Clinton he felt competitive with. Still, he and Hillary often had lunch together in the vice presidential residence and had developed a rapport.

Obama was a friend to Biden and listened to his deliberations but did not weigh in, said Valerie Jarrett, who was a senior adviser and remains a close confidante to the Obamas. "President Obama knew that if he made the decision [to run for president] because someone else told him to, that that is not the right way to make that decision. The only person who could have talked President Obama out of running for president was Mrs. Obama. He had to decide whether the fire was in his belly."

Beau's loss was, understandably, all-consuming. "The family was broken, and I was more broken than I thought I was," Biden has said. In a speech at Colgate University four months after the 2016 election, Biden said, "Do I regret not being president? Yes. Do I regret not running for president, in light of what was going on in my life at the time? No."

Being upstaged by Hillary Clinton was nothing new for Biden. His chief of staff Ron Klain, who had been Al Gore's chief of staff, says he aggressively encouraged Biden to make his intentions known early, to be his own man and cultivate his own relationship with donors—Klain knew firsthand just how tricky positioning a vice president for the presidency could be. But in 2012, while on a fund-raising trip to California for Obama's reelection, when word got back to David Plouffe and Jim Messina, who were running Obama's reelection campaign, that Biden was trying to add meetings with donors from Hollywood and Silicon Valley for his own purposes, he was summarily told to stop it—immediately. Plouffe took Alan Hoffman, Biden's deputy chief of staff, into his office and made it clear that everyone had to have a singular goal in mind: getting Obama reelected. "They came down hard on him [Biden]," said a former top Biden staffer. Messina and Plouffe interpreted Biden's efforts as setting himself up for 2016 instead of focusing on the race at hand. Messina was particularly enraged by it, two members of Biden's staff said. "Messina's a kamikaze," one staffer said. "He's either furious or invisible." It was a missed opportunity. Klain, who went on to work for Hillary Clinton's presidential campaign, said that he wanted Biden to start positioning himself to run for president as early as 2010.

But Biden's position was more precarious than he knew. To add insult to injury, Biden did not find out that the Obama campaign was conducting polling in late 2011 to see if he should be replaced as vice president with Hillary Clinton until he saw reports in the press (the poll results showed she would not be much help and the switch was ruled out). The president dismissed the polling and said it was something the staff did on their own. But it hurt Biden, who had left a job he loved in the Senate to be Obama's vice president and thought he was doing a good job. Christopher Hill, who had served as U.S. ambassador to Iraq, said he talked with Biden at the time and Biden was furious. Biden told him: "This idea that she would help the ticket if Barack replaced me with her, we looked at all this in 2008 and there were a lot of focus groups and polling data about this and it showed consistently that she could not carry Pennsylvania, Ohio, or even Michigan." The campaign could have done a better job of pouring cold water on the story, said Biden's former chief of staff Bruce Reed. He added, "I'm sure the only person less interested in that scenario than Joe Biden was Hillary Clinton." Hillary, everyone knew, wanted to be president, not vice president.

When he considered seeking the presidency for a third time in 2015, he was well aware that he had not run in a competitive race since 1972, when he beat Republican senator Cale Boggs by a slim margin (Boggs had not lost a race since 1946) and replaced him as senator. He recognized that it would be an uphill battle. But when he tried to get Wall Street financiers such as Robert Wolf, who had been a strong supporter of Obama's and a friend of Biden's, he found out he was too late: Clinton had already locked Wolf down. Although he was friendly with Clinton, Biden knew that her campaign operation would come after him if he decided to run. Hacked emails of Clinton campaign chairman John Podesta show that when Biden was weighing a White House run there was panic inside the campaign. Steve Elmendorf, a longtime Clinton supporter, emailed the campaign to make sure it was aware of the anxiety within the party about a potential Biden run. "I get multiple freak-out calls every morning and I try to talk everyone off the ledge and not bug u all," Elmendorf wrote.

Then the attacks began. The *New York Times* published a story— "Joe Biden's Role in '90s Crime Law Could Haunt Any Presidential Bid"—on August 21, 2015, citing controversial anticrime legislation Biden helped pass in the 1990s that critics argue led to mass incarceration. One close Biden ally questioned the timing of the article's publication and the source of the story. "I'll bet you ten to one it was somebody on the Hill who worked on the crime bill and worked for Hillary's campaign." Another Democrat who is close to Biden and Obama said, "The Clintons can be a little bit heavy-handed when it comes to making sure that competition goes away." Biden was hurt by the headlines. "He has these old-school Marquess of Queensberry rules [a code of rules in boxing used to refer to fair play] that Democrats don't do that to each other," said Tony Blinken, who was Biden's national security adviser and who later worked for Obama. "I think there was an element of being personally offended by it." The Clintons had an enormous campaign apparatus already in place, and a Biden challenge, some Clinton allies later argued, could have led to Vermont senator Bernie Sanders winning the nomination.

One top Biden aide said that his staff "worked hard to reposition him in the moment as opposed to the past." The aviator glasses and the "Uncle Joe" nickname did not bother him. "In fact," the aide said, "seeing him as a cool older guy who wears a bomber jacket is much better than being a gasbag in the Anita Hill hearing." Biden has faced long-standing criticisms for his role in the 1991 confirmation hearings for U.S. Supreme Court nominee Clarence Thomas, which he helped conduct as the chairman of the Senate Judiciary Committee at the time. Biden was one of fourteen white men on the judiciary panel and has since apologized for not doing more to defend Hill during the hearing where she was asked deeply personal questions about her sex life. If Biden had jumped into the race in 2015, it would have been hardball and he could not have avoided addressing it. He has since said he owes Hill an apology.

But Beau's illness and passing was the biggest influence. Time for Biden to jump in had simply run out, so he turned his attention to helping Clinton's campaign. Biden used his working-class roots on the stump as he traveled around the country to support Clinton. He would get briefings from Clinton campaign manager Robby Mook and from Podesta. But while everyone around her thought she would win given the polling, Biden was not as optimistic about her chances of beating Donald Trump, the Republican nominee. He never thought people were excited about her, and he told senior campaign staffers that Clinton should focus on what she could do to help the middle class and not just criticize Trump.

Blinken was on the phone with Biden five or six times the night of the 2016 election. Biden was watching the results in the Observatory. "It was starting to go south and when I first called, Biden said, 'It's OK, we'll be fine,'" Blinken recalled. "Then a little later he said, 'There's still a path.' And then, ultimately, it was, 'I can't talk.'"

Biden was immediately worried about the foreign policy implications of Trump's victory. National security, he thought, would no longer be in safe hands. After the election Biden was consumed with guilt, according to friends. He thought that if he had run, he would have won. And if he had been able to convince Clinton's campaign to focus more on what she could do to help the middle class rather than focusing on attacks against Trump, maybe Clinton could have won.

Since leaving the White House, Biden is working for both the University of Pennsylvania and the University of Delaware. At Penn he heads a center that focuses on diplomacy and national security and at the University of Delaware he leads an institute that centers around public policy solutions to issues including criminal justice and women's rights. And he is slowly and methodically laying the groundwork for a 2020 presidential run. He is surrounded by his longtime political aides Ted Kaufman, Steve Ricchetti, and Mike Donilon, and he has set up a PAC—the American Possibilities Political Action Committee. He has survived two life-threatening brain aneurysms, and if he runs in 2020 he will be seventy-seven years old on Election Day—that's seven years older than Donald Trump was when he became the oldest president to take office. Because of his health history and his age, Biden is considering making an unprecedented campaign promise: he might run for one term only.

————

Joseph Robinette Biden Jr. was born in 1942 in blue-collar Scranton, Pennsylvania, the son of a car salesman and a homemaker. He was one of four children raised in a loving Irish-Catholic family (he later became the first Catholic vice president). When he was ten years old his family moved to the Wilmington, Delaware, area, where his father found work after struggling to find a job in Scranton. Biden often talks about his roots in Scranton and sitting in his "Grandpop Finnegan's" kitchen listening to relatives argue about President Dwight Eisenhower. As a kid he had a stutter and was nicknamed "Joe Impedimenta" by his classmates. One classmate called him "Dash" for "J-J-J-Joe Biden." His mother, Jean, encouraged him to stand up to bullies and told all her children: "You're not better than anybody, but nobody is better than you." Biden's uncle Ed stuttered, too, but was never able to conquer it. He never married, never got a good job, and he drank a lot. Biden never wanted to end up like Uncle Ed, so he never had a drink and he never stopped working to get over his stutter, memorizing poems and practicing speeches endlessly.

He graduated from the University of Delaware and got his law degree from Syracuse University. In 1966 he married Neilia Hunter, and they had three children. Inspired by the civil rights movement of the 1960s, Biden got involved in local Democratic politics in Delaware. "I wasn't at the bridge at Selma, but the struggle for civil rights was the animating political element of my life," he said. When he ran for Senate in 1972, Biden campaigned on his support for integration.

He became the fifth-youngest senator in U.S. history when he was elected in 1972 at the age of twenty-nine. He seemed to have a charmed life until, one week before Christmas and days before he was set to be sworn in, Neilia and their thirteen-month-old daughter, Naomi, were killed in a car crash when their station wagon was hit by a tractor trailer. Neilia was on her way to buy a Christmas tree with their young children and Biden was in Washington setting up his office and hiring staff when he got the call. "You knew when the call came. You knew," Biden told families of fallen soldiers in a moving 2012 speech at a TAPS National Military Survivor Seminar. "You just felt it in your bones something bad happened." His two sons, Beau, then three, and Hunter, then two, were also in the car and were critically injured. They spent weeks recuperating in the hospital, and eighteen days after the accident that killed his wife and daughter Biden was sworn in at his sons' bedsides. "For the first time in my life," Biden said, "I understood how someone could consciously decide to commit suicide, not because they were deranged, not because they were nuts, because they had been to the top of the mountain and they just knew in their heart they would never get there again, that it was never going to be that way ever again."

He considered suicide in those earliest and darkest days, but he couldn't leave his two young sons. His sister, Val, who Biden calls his best friend, moved in to help take care of the boys, and the Senate became a second family to him. He began a near daily three-hour round-trip, thirty-six-year-long ritual, taking an Amtrak train between Washington, D.C., and his home in Wilmington so that he could be home with his sons. They became his salvation. "Looking back on it, the truth be told, the real reason I went home every night was that I needed my children more than they needed me," he said, noting that some commentators said his weekend absences from Washington—he is nicknamed Amtrak Joe—would damage his career.

"If you ask me who's the unluckiest person I know personally, who's had just terrible things happen to him, I'd say Joe Biden," said Ted Kaufman, who, next to his sister Val, is Biden's closest friend and was his chief of staff in the Senate for nineteen years. "If you asked me who is the luckiest person I know personally, who's had things happen to him that are just absolutely incredible, I'd say Joe Biden."

Biden spent thirty-six years in the Senate and has served as chair of the judiciary committee and the foreign relations committee. In 2008, one of those "incredible" things happened when Obama picked Biden as his running mate. Biden thought it was like joining a team about to go into the Super Bowl as its running back. His job was to just carry the ball and not be engaged in strategy. "Can you imagine how hard that was? At his age?" Kaufman said.

Biden walked into Obama's campaign office in Chicago, where hundreds of people were working for Obama, with just about a dozen aides. "It's kind of awkward," said Alyssa Mastromonaco, who was the director of scheduling and advance on Obama's campaign. Biden's first exposure to Obama's always-on-message aides during those early days on the campaign, and in the White House, proved difficult.

After the 2008 election, Biden spoke with Delaware's Democratic governor Ruth Ann Minner about who would fill his Senate seat during the last two years of his term, a post he'd held since Obama was just eleven years old. Biden was pushing for Kaufman, his former chief of staff and a Delaware resident, to take his seat. "Have you talked to the president about this?" an Obama staff member asked Biden. "Why the hell do I have to talk to the president about this? It's my state, it's my friend," Biden replied incensed. But he was in a new position now, he had a boss and everything a vice president does reflects on the president and all the power a vice president has is derived from the president. The president-elect probably wouldn't care, but he needed to be notified. Kaufman was eventually appointed to fill Biden's Senate seat.

It was the beginning of an education for Biden, who was initially uncomfortable working for someone after more than three decades in the Senate. And it was a reality for his entire eight years as vice president: when he wanted to bring in Kevin Sheekey, who was New York mayor Mike Bloomberg's top political adviser, as his chief of staff after Ron Klain left, Obama advisers Plouffe and Axelrod said no—they did not think Sheekey would be a good fit. Even Obama said he did not think Sheekey would work out. He did not pass their all-important loyalty test and Biden had no choice in the matter.

When Obama, who is almost two decades younger than Biden, came to the Senate in 2005 he was already a superstar, having given a rousing speech as an Illinois legislator at the Democratic National Convention in 2004. "To people who had been around a long time, that rubbed some of them the wrong way because they had been there for years and paid their dues and here's this person who's being talked about for the presidency," said Blinken, who was Biden's staff director on the foreign relations committee before following him into the administration. Obama, Biden thought, was in too much of a hurry.

When Biden was running for president against Obama in the 2008 primaries, he would sometimes say, "There's no way that the Democratic Party is ever going to nominate someone who is a first-term senator, after 9/11, his name is Barack Hussein Obama, it's not going to happen." There was an arrogance there that bothered Biden. "We used to be on the tarmac for campaign events and Biden and [Connecticut senator Chris Dodd, who also ran in 2008] were sharing a plane from Iowa to go back in time to vote and they were looking at each other, 'What's wrong with this picture? We're carrying our own bags.'" Biden is in many ways the opposite of Obama—he thrives on interaction with people, while Obama is self-contained and happy to be on his own. Aides sometimes had to remind Obama to try to remember the names of people's dogs and grandchildren, something that comes as second nature to Biden.

Biden and Obama did not know each other well in the Senate, even though Obama was on the foreign relations committee when Biden was its chairman. In 2007, on the same day that he announced he was running for president, Biden found himself defending remarks he made to the *New York Observer* in which he weighed in on his rivals for the Democratic nomination. "I mean, you got the first mainstream African American who is articulate and bright and clean and a nice-looking guy," he said. "I mean, that's a storybook, man." Biden did not immediately understand what all the fuss was about—an aide had to call and tell him why the remark was a problem.

A few days later, Biden was chairing a foreign relations committee hearing and Obama was running late. Obama wanted to ask a question and his staffer kept asking Biden's staff to keep the hearing going. A note was passed to Biden imploring him to not end the hearing before Obama arrived, so Biden stalled by asking more questions. When Obama walked in, he turned to Biden's aide and said with a smile, "Boy, Joe's being awful nice to me lately." Just two years earlier, at his first meeting of the Senate foreign relations committee in 2005, which happened to be a confirmation hearing for Condoleezza Rice, Obama passed an aide a note in the middle of one of Biden's long monologues. It read: "SHOOT. ME. NOW."

Once he decided to sign on as Obama's running mate, Biden had to come to terms with Obama as "the chosen one," and Obama had to come to terms with his own lack of experience. "Both of them ate a certain amount of shit to see the virtue in the other person," said a former Obama administration official. Obama, always analytical, recognized that Biden's deep experience on foreign policy and national security issues and his relationships with world leaders and with members of Congress would augment Obama's own strengths and would fill in some of his gaps.

The Obama campaign said Biden could not have any of his own aides on the campaign plane when he was traveling, but Biden shot back and said, "If Mike Donilon or Ted Kaufman aren't on the plane, I'm not getting on the plane." The campaign relented. Every morning Biden, who is so loquacious that aides were constantly worried that he would put his foot in his mouth and who they sometimes treated like a rambunctious child, would get on a conference call with Axelrod and Plouffe, who would tell him: "Here's where you're going and here's what you're saying."

During the campaign, Biden made several gaffes, including one at a Seattle fund-raiser when he told a room full of donors, "Watch, we're going to have an international crisis, a generated crisis, to test the mettle of this guy. Mark my words. It will not be six months before the world tests Barack Obama like they did John Kennedy. The world is looking. We're about to elect a brilliant forty-seven-year-old senator president of the United States of America."

Fund-raising events are generally closed to the media, but Biden apparently forgot that that night's fund-raiser was open to reporters. He declared, "I probably shouldn't have said all this because it dawned on me that the press is here." Moments later, Biden ended his remarks. "Obama was irritated about that," Axelrod said. Biden is still unapologetic about the slips. "I would say some things that were obvious, like he's going to be challenged in the first year, like that was a big goddamn secret—every president's challenged in the first year," he said, still surprised that it upset anyone.

During a meeting in September 2008, Biden told Obama that he would need to have access to all the information Obama had. Biden was on his way to Minnesota to campaign for Senate candidate Al Franken, and he told Obama, according to an internal memo, that he "would need to have known what BO and the campaign had communicated to and about Franken" so that he would know what to say. If everything was going to be so carefully calibrated, and if all his words were going to be parsed, then he would have to be included in everything. "All presidents pick vice presidents for political reasons, but these relationships are not sometimes what they appear to be," said one Biden aide. "Early on it was not a very good relationship between the VP and the president." Obama and his close cadre of advisers saw it as their game and they had worked hard to win the nomination. Biden was along for the ride.

Obama told Biden he had veto rights over anybody in his Cabinet. "A number of the Cabinet members I knew better than the president knew them because I'd worked with them for a long time," Biden said. He likes to point out that his aides, Klain and Kaufman, sat with the Obama team and made decisions together. "At least half of the president's staff were my former staff members," Biden said. "It [the relationship] rests on one thing: defining what it is you want to do and determining whether or not the president-elect agrees with you."

Biden told Obama he wanted two things: "To be able to completely be level with you and argue with you if we disagree privately, and secondly I want to be the last person in the room on every major decision. I didn't mean it figuratively, I meant literally." Obama agreed to the terms and said, "I want to make clear that when I have a disagreement with you I'm going to flat out tell you and I want you to do the same with me and do it in private." Biden agreed and said he would resign before he disagreed publicly with Obama: "If we ever get to a place where I have a fundamental moral disagreement with you," Biden said, "I'll develop [tell people I have] prostate cancer and resign. I will not publicly take issue with you on anything." He saved disagreements for their lunches, or he would stay in the Oval Office after other advisers left to talk to Obama in private.

Biden has called the Bush–Cheney relationship more "codependent" than his relationship with Obama. Biden recommended retired Marine Corps general Jim Jones to be Obama's first national security adviser. Jones was initially concerned that Biden would want to have a shadow national security council, like Cheney had, because of his years on the Senate foreign relations committee. Biden told Jones, "I'm not going to keep that staff at all on one condition, and you should talk to the president about this: I have total access to the national security council." Jones replied, "Thank God."

Biden said he spent four to seven hours a day, every single day, with the president, and he was in every major meeting, at Obama's request. At 7:40 every morning his senior staff would meet with Obama's senior staff to set the day's agenda. Biden said when he was assigned a job, like overseeing the 2009 Recovery Act, a nearly $1 trillion stimulus package meant to revive the economy after the financial crisis, he had complete authority over it. One Cabinet member disagreed with him in an early meeting and Biden said he replied with a not-so-veiled threat: "Look, it's simple. Do exactly what the hell I tell you, or go see the president. I don't mind. Well, son of a bitch, we had no problems after that."

Even toward the end of their time together, there were moments when Obama grew exasperated with Biden. During one meeting about Iran, Biden started to discuss the Sunni/Shia divide and Obama cut him off in front of everyone in the room. But, ultimately, they had a partnership that did something no other modern president and vice president have managed to do—it grew stronger over time, personally and professionally. "That 'bromance' thing is real, man," Biden says. And there was mutual reverence and respect between them—Biden often tears up when he talks about his relationship with Obama and Obama speaks affectionately about their friendship. "I'm grateful every day that you've got such a big heart, and a big soul, and those broad shoulders," Obama told Biden at Beau Biden's funeral. "I couldn't admire you more."

"In the way a good friendship is, they were honest with each other, sometimes they had disagreements that could get to the point of being intense," said Biden's former chief of staff Steve Ricchetti. "They were close enough that they could say things strongly to each other and still were brothers afterward . . . They willed it to work."

At the Democratic National Convention in 2008 Biden's grand-daughter Finnegan asked if she and her sister Maisy could have a sleepover at the hotel with the Obamas' daughters Sasha and Malia and their cousins. "And so I went to Jill and Jill called Michelle and Michelle was already on it," Biden said. The hotel prepared a room next to theirs for the girls. Like any kids they slept on futons, watched TV, and ate popcorn. "I remember walking out to go to the floor [of the convention] and opening the door and seeing all these little black and white faces lying down with each other, some in the same sleeping bag, and I knew, I knew this was going to work."

About a year into the administration, as they were grappling with the economic crisis and wars in Iraq and Afghanistan, Biden called David Axelrod into his office and shut the door. "Do you remember that conversation we had when you came out to see me in Wilmington?"

"Sure I do," Axelrod replied.

"Remember when I said I thought I'd be the better president?"

"Yeah."

"I was wrong about that," Biden said. "I'm so proud of this guy and I'm so proud to be his partner, and I just wanted you to know that."

———

Biden was the first person in his family to go to college. He has always been defensive about not having attended an Ivy League school, like so many of his Senate colleagues. Before Senator Russ Feingold, a Democrat from Wisconsin who happened to graduate from Harvard Law School, showed up for his first hearing as a member of the Senate judiciary committee, he asked an aide to fill him in on the chair of the committee, Joe Biden. "One thing you need to know about Joe Biden is he has a chip on his shoulder about where he went to school," the staffer told Feingold, who was surprised that after so many years in the Senate Biden would still harbor those feelings. After Biden gaveled the hearing into session, he turned to Feingold and said, "Welcome to the committee. While I don't have the pedigree and didn't go to the schools you went to . . ." Feingold looked at his aide and smiled. Biden knew the jokes about him and that some of his critics thought he was not sophisticated enough. Dick Cheney confirmed Biden's paranoia. When asked for his take on his successor he called him "a good guy, a nice guy," but he added: "Personally we always got along fine . . . I don't see him as a heavyweight." (Biden for his part called Cheney "the most dangerous vice president in U.S. history," but since Trump's election he has been far less critical. "Give me back the neocons," he said. "Cheney was really smart and really informed. I think he was wrong, but not because he wasn't informed.")

Obama's campaign team would worry about "Joe Bombs," as they called them, especially in the early days. "Uncle Joe" was gaffe-prone but lovable. "It [speaking off the cuff] has held him back in his career," said one former aide. "He was a goofball, his reputation among reporters was terrible." During one particularly cringe-worthy moment in the 2008 campaign at a televised event he said: "Stand up, Chuck! Let 'em see you!" to Missouri state senator Chuck Graham before realizing he was confined to a wheelchair. In February 2009, after the Inauguration, Biden told an audience that there was a "thirty-percent chance we're going to get it wrong" on the economy. When a reporter asked Obama about it he said, "I don't remember exactly what Joe was referring to. Not surprisingly." In March 2010, when Obama signed the Affordable Care Act in the packed East Room of the White House, marking what Democrats consider to be the biggest legislative accomplishment of his presidency, Biden famously whispered into the president's ear that it was a "big fucking deal." The off-the-cuff remark was picked up by a live mic and created a distracting story line for the West Wing.

The "Uncle Joe" memes and GIFs, the most famous being a shirtless Biden washing a Trans Am parked in the White House driveway by the satirical *Onion*, have made Biden into something of a sensation. And his Ray-Ban aviators and leather bomber jacket added to his allure. He was perfectly happy running around the lawn of the Naval Observatory for an annual summer beach party with White House reporters with a Super Soaker spraying everyone in sight. Yet his public persona masked a surprisingly sharp temper that he quickly turned on members of his staff. Aides refer to this informal hazing as being "Bidened" or "Bidenized." "He'd blow up at you," said one staffer. "It was terrifying the first time," said another former staffer who has a good relationship with Biden and overall enjoyed working for him. "He'd grit his teeth and say, 'Why the fuck do I have to do this trip?'" He has been known to make the uninitiated cry, and it can be difficult for speechwriters to work for him. Sometimes he barely read their work before rewriting it. Being a speechwriter for Biden, aides joked, was the worst job in the White House. But Biden speechwriter Dylan Loewe marveled at Biden's Socratic approach. In meetings Biden drilled down tirelessly on an issue, and in some ways he was more demanding to work for than Obama. Biden, his aides said, expected briefings that were worthy of the office he occupied. "You only got yelled at," Loewe said, "if you didn't meet the mark." The job of working for the vice president, Loewe pointed out, "is not supposed to be easy."

Loewe worked for Biden during his final two years in office and says that Biden never completely overcame his stutter. "You never totally overcome a stutter. There were times when his stutter would appear in meetings." On the way to a speaking event, even though he was going to read the speech on the teleprompter, Biden would look it over and do what he had done as a kid and divide the words into musical beats. "Almost like he was a guitarist about to go out and sing a new song," Loewe recalled. "He wanted to make sure he had the rhythm right." What Biden's speechwriters were trying to do, Loewe said, "was more than just write a speech, we were trying to get the words right so that they wouldn't cause the muscles in his mouth to spasm." The irony was not lost on Loewe: "Here's a guy that most people think of as loquacious, and he is, but words are not always easy to get out of his mouth."

Halfway through speech prep in Biden's West Wing office one day, his secretary came in with a note. Biden interrupted the discussion and went to the anteroom and brought in a mother and her young elementary-school-age son. Biden had met them a couple of months earlier in a photo line, and the woman had told him how inspiring he was to her son, who also has a stutter. Biden had given her his phone number and told her to visit if she was ever in Washington. "You're way smarter than I was when I was your age and I'm vice president now," he told the young boy. "I want you to watch our speechwriting so you can see, if I can do this, you can do this."

At a rally in New Hampshire during the 2012 campaign, Biden was to introduce the governor, and the governor would introduce President Obama. Loewe wrote a three-minute speech. When it was over, Biden had not changed a single word and kept to the time limit. Offstage, Loewe and another staffer high-fived each other—they were so amazed. President Obama, who was seated nearby, looked bemused. "Wait," he said, "Joe's done already?"

———

On December 14, 2012, twenty-year-old Adam Lanza opened fire with a military-style assault rifle at Sandy Hook Elementary School in Newtown, Connecticut, killing twenty first-graders and six adults. Obama called it "one of the worst days of my presidency," a tragedy that brought tears to the eyes of normally stone-faced Secret Service agents. Obama tapped Biden to lead the push for gun control legislation and to engage with the Newtown families. Biden was able to talk to the grief-stricken parents in a way that others could not. Having lost his wife and daughter, he was able to empathize. "It doesn't come as naturally to the president," Biden's chief of staff Bruce Reed said. "With victims Biden has a special rapport because he knows personally what it's like, and he's also a guy who wears his emotions on his sleeve."

Shortly after the tragedy, some of the parents went to the vice president's large, high-ceilinged ceremonial office in the Eisenhower Executive Office Building. They sat at the long table clutching framed photographs of their children, some holding their child's favorite stuffed animal; others wore the names of their children on wristbands. Some parents brought the siblings of their murdered children, most not more than ten years old themselves. They went around the long oval table in the ornately decorated room with mahogany floors, chandeliers, and two fireplaces. As parents told stories about their children, other parents and White House aides reached for the boxes of tissues lining the table. "I've been to a lot of sessions where the president or vice president had to deliver bad news or grieve with families, and this was the most genuine heart-wrenching one I've ever seen," said Biden adviser Sarah Bianchi. The meeting was so emotional, Reed said, that he told other staffers they could leave if they needed to. Biden, with tears in his eyes, told the gathered parents how much he admired them, how brave they were to try to do what they were doing to change gun laws and turn their personal suffering into something positive.

Biden confessed to the families that he had always felt like a coward after the accident that killed his wife and daughter, how a tractor trailer was involved and how he had an opportunity to push for tougher regulation on tractor trailers, which he did not do in the Senate. "You have more courage than I did," he said, as tears welled in the eyes of many people sitting at the table. He told the family members how devastated he had been after the accident. He said the best advice he got then was to keep a calendar and ask at the end of each day whether that day was better than the day before. For a long time, he told them, it will not be better, but after a while they would start having more OK days than bad days.

The most gut-wrenching meeting came when the Senate voted down a bill that would have extended background checks for firearms purchases and family members came to the White House for a Rose Garden statement. During one meeting with members of the NRA, Biden was angry and arguing, his voiced shaking with indignation. "The vice president knew from the moment he got the assignment to shepherd the legislation that we weren't going to be able to get anything done apart from what we could do with executive action," Reed said. Before the event in April 2013 Obama and Biden gathered in the Roosevelt Room with family members and that meeting, Reed said, was even harder than the first because now the only thing the families had left to do was grieve.

After the Pulse nightclub shooting in Orlando, Florida, in the summer of 2016, when forty-nine people were killed and dozens more injured in a terrorist attack, Biden insisted on accompanying Obama to a memorial service there, even though some Obama aides thought it was best for the president to go alone. Biden was adamant about going to meet with the families because he could relate to their pain, having recently lost his son. Obama and Biden stood about twenty feet apart as they greeted the victims' families. Valerie Jarrett stood closer to Biden, and she could hear him tell family members about the loss of his son Beau and the loss of his wife and daughter. "I know what you're going through," he told them. "I just want to assure you that it does get better. Your tears of sorrow will turn to tears of joy [when you remember your loved one], but it takes time."

Jarrett marveled at how emotionally vulnerable he was and how that allowed them to confide their grief to him. "It allows people to open up to him with a level of intimacy that I've rarely observed with strangers," she said. Biden is very physical, he will grab you and hug you and pull you close to him. Jarrett said that physicality is "motivated by his desire to connect." People, she said, "feel it, it's palpable, it's uncomfortable sometimes because most people aren't that comfortable showing their emotions." Biden, she said with admiration, is "fearless about emotion."

———————

Biden says he helped Obama be more demonstrative with his own emotions and trust his instincts. Biden said that every decision in the White House went through a rigorous process before it ever got to Obama. Before a national security issue ever reached his desk it went to deputies at the National Security Council, then to top aides, and finally to Obama. "I used to say, 'Mr. President no matter how hard you work, no matter what you do, no president is ever going to have more than sixty to seventy percent of the facts upon which to make a critical decision. Never. Does not exist. So your instinct *really* matters." Obama, Biden said, "was constantly looking for more information, more data." Biden saw it as his job to get him to listen to his gut.

The best example of that dynamic is the May 2, 2011, Navy SEAL raid that killed Osama bin Laden, the al-Qaeda leader behind the 9/11 terror attacks, who was hiding in a compound in Abbottabad, Pakistan. "He knew his presidency was on the line with bin Laden," Biden said. "Think about if that had failed, his presidency would have come to practically a screeching end." On the night of April 28, Obama gathered top members of his national security team, including Biden, Defense Secretary Robert Gates, CIA director Leon Panetta, chairman of the joint chiefs of staff

Admiral Mike Mullen, and Secretary of State Hillary Clinton in the Situation Room and essentially did a roll call around the table, asking each whether they thought he should give the order to send the SEALs in to kill the fugitive al-Qaeda leader. Biden came down against going in—without being absolutely sure that bin Laden was in the compound, Biden argued, more time was needed to get a positive identification. Most people were hedging their bets, saying they were 49/51 whether he should authorize the raid or not. "Only three people said unequivocally go or don't go: two that said unequivocally go were the chairman of the joint chiefs and Panetta. One who said unequivocally don't go was the secretary of defense. I'm the last guy in the room and it comes to me. I turned to everybody and said, 'You've got seventeen one-handed economists in here [Harry Truman famously complained, "All my economists say, 'on the one hand . . . on the other'"]. That's wrong, you have fourteen [three people actually gave firm opinions]. Mr. President, I think you should choose option two [not to go]."

During the 2016 election, Clinton, who was "51/49 to go in" according to someone in the room, said she was unequivocally for the raid. In January 2016 she told an audience at Iowa State University: "I was one of those who recommended the president launch what was a very risky raid . . . because if all we had done was launch a missile and dropped a bomb we never would have known [if bin Laden was dead]." Obama's top strategist David Axelrod said, "My sense is that she was not sold on the idea either." Biden clearly resents the way Clinton mischaracterized her position. One aide said: "The ass covering, opportunistic version really rattled him." Biden advised against the raid, he said, "to give room, because then I walked out with him [Obama], which I always did, and walked to the office. I said, 'Follow your instinct. I think you should go, follow your instinct.' I could tell that was his instinct." Biden said he fell on his sword so that Obama would look like he made the call on his own. "I wanted people to know what a chance this guy took. If I had said that I said to go [at the time] then it would make me look like I was ratting out everybody else." Press reports made it seem like Biden was one of the lone voices against what ended up being a successful raid that won Obama widespread praise. "That's the job of the vice president," Biden said. "You're supposed to throw yourself in front of the train. That's one of the reasons I didn't want to do it [become vice president] in the beginning."

Biden was reminded again and again that his legacy is attached to the president's. "In order to be a really successful vice president, you have to subvert your own interests," Biden's former communications director Shailagh Murray said. "The most important thing was being the best vice president you could possibly be and sometimes that was going to require eating shit." That was the case when Biden appeared on *Meet the Press* during the reelection campaign in May 2012, and he declared his support for same-sex marriage. "I am absolutely comfortable with the fact that men marrying men, women marrying women, and heterosexual men and women marrying another are entitled to the same exact rights, all the civil rights, all the civil liberties," Biden told host David Gregory. "And quite frankly, I don't see much of a distinction beyond that." Biden had recently met with a group of gay Democrats in Los Angeles and the subject had been on his mind. He knew that the president's hedging—Obama said his opinion on the matter was "evolving"— was hurting the party. He did not think he had gotten in front of the president until he read the interview transcript himself.

Before appearing on Sunday shows Biden would get briefed on foreign and domestic policy—sometimes the domestic policy briefings could last more than an hour. He was relentless about these briefings, sometimes agonizingly so for his staff. But gay marriage had not come up. Biden adviser Sarah Bianchi stood in the studio during the interview but could not clearly hear his answer on gay marriage.

When he came off the set he asked, "I got all the economic stuff right, right?"

"Yeah, you did great. What was that you said on gay marriage?"

"I said the president makes policy on love." But that was, of course, not what he said.

"I'm confident it was not strategic," said Murray, who was also with him at the interview. "But once he did realize what he had said, his attitude was: might as well embrace it. He thought he could be the guy who could throw himself on the hand grenade and let the chips fall where they may."

Obama's top campaign aides were furious—this was not part of the long-term plan and they thought Biden was being disloyal. And it got the White House off what was supposed to be its key message that week: economic issues and student loans. Biden speaking out on same-sex marriage made the president look weak, and it forced Obama to confront the controversial subject head-on.

Obama's campaign manager Jim Messina, in particular, was extremely upset and called Biden aides screaming about the vice president overstepping and forcing Obama into a corner during a hard-fought election. "They were the control freaks, and that's the way campaigns are supposed to be," said one Biden staffer. "It took something away from the president which Biden was cognizant of so he reached out to him and apologized for putting him in that position. He regretted that but he did not regret what he said." The situation was handled clumsily, and aides say that was solely Biden's fault. But Biden's aides thought the vitriol coming from campaign headquarters in Chicago was out of proportion to the crime. "I knew exactly what they were doing," said one Biden aide. "I told those guys—Messina in particular—to shut the fuck up and stop talking to reporters. That we would end the story with an apology."

Obama was blindsided by Biden's remarks, but he was the least upset, and once Biden personally apologized to Obama everything was forgiven. But Biden's staff was shaken. Obama said at a staff meeting the next day that Biden was "speaking from the heart" and it was fine. Obama told Axelrod, "I can't punish Joe for being bighearted. I'll talk to him about discipline, but I know where that came from, I know why he answered the way he did. He wasn't trying to upstage me." On May 9, three days after Biden's interview, Obama sat for an interview with ABC's Robin Roberts. According to one account, Michelle Obama considered it a blessing in disguise and told her husband before his interview: "Enjoy the day. You are free."

Obama told Roberts, "I've just concluded that for me personally it is important for me to go ahead and affirm that I think same-sex couples should be able to get married."

But the lead-up to the ABC interview was difficult for Biden's top advisers, who were in the firing line. "The West Wing staff was just horrible about the whole thing and spent the two days torturing us and resenting that they had to deal with it now," said a former Biden aide. "It's not always direct abuse. Sometimes it's leaking to the press, trash talking. It was a classic example of the clash between the no drama control freaks and the 'Look mom, no hands!' approach." The campaign insisted that the president had had a plan to address the issue and Biden had forced them to come out with it sooner than they would have liked. It was one of several periods when Biden and his team were put on probation. Though it was a major setback for him internally, it won him lots of support from people who praised him for speaking from the heart, even when it was not politically expedient.

Less than six months later, Biden redeemed himself during the October 11, 2012, debate with Republican nominee Mitt Romney's running mate, Wisconsin congressman Paul Ryan. The debate came at a precarious time for Obama after his listless first presidential debate with Romney eight days earlier. There was a lot of hand wringing inside the White House and a sigh of relief when it was over. Even critics agreed Biden had won the debate. Obama watched it on board Air Force One, and when it was over he called Biden to congratulate him on his performance. Though there was no plan to speak to reporters when he arrived in Washington, Obama felt compelled to talk to the press gathered on the tarmac at Andrews Air Force Base. "I'm going to make a special point of saying that I thought Joe Biden was terrific tonight," Obama said. "I could not be prouder of him. I thought he made a very strong case. I really think that his passion for making sure that the economy grows for the middle class came through. So I'm very proud of him." It was a critical moment for Biden, who had been in hot water months before. "It was important to the President that he show publicly and privately that he had the Vice President's back," recalled Obama's press secretary Josh Earnest.

Biden's last chief of staff, Bruce Reed, said the West Wing always worried about Biden breaking new ground, but that was what made Biden a good teammate for an unusually reserved president. "We worried more about Biden getting his dates mixed up, there are all kinds of mistakes you can make in an interview," Reed said. "A lot of politicians who are very careful in what they say are less likely to make mistakes, but they're also less likely to say anything great." But in Biden's case there was an effort to make sure he did not say anything that ran counter to the administration's message, especially after *Meet the Press*.

During the 2012 reelection campaign Obama expanded his tight inner circle to include Biden and his aide Mike Donilon. He wanted more voices at the table, but expanding the circle came at a price. During their first strategy meeting in the White House's State Dining Room, Obama was very candid about what he had failed to get done in his first term, including his promise to close the U.S. military detention facility at Guantánamo Bay in Cuba. When details of the meeting emerged, Obama, who notoriously hated leaks, was furious. He called in all the attendees of that first meeting and chastised the entire group: "I expanded the circle, I put my faith in this group and someone let me down. I expect that person to come and tell me about it." He walked to the Oval Office and no one followed him. Biden stood up and said: "This guy would do anything for you. He trusted you." Then Biden left the room. But Biden and his team came under suspicion for being the source of the leak, so Biden and Donilon were cast out of the fold and no longer invited to strategy meetings.

Still, Biden was a useful asset for the West Wing. At the end of 2012, after Obama won reelection, the Bush-era tax cuts were set to expire. The administration wanted to extend the middle-class tax cuts and end the tax cuts for the wealthy. But congressional Republicans threatened to shut down the government if deep budget cuts were not enacted to pay for the middle-class tax cuts. If no deal was reached there would have been a massive tax increase on the middle class. Biden wanted to help but the West Wing did not want to let him anywhere near the issue and disinvited Reed and Bianchi from meetings at different junctures. Obama had promised Senate Majority Leader Harry Reid that he would not involve Biden in negotiations with the Senate's top Republican, Mitch McConnell, because Biden had a reputation among Democrats for giving away too much in negotiations.

Eventually it was McConnell and Biden, however, who would reach a deal. During one near-crisis, McConnell called Biden and asked rhetorically, "Does anyone down there know how to make a deal?" In negotiations, Biden's first question is *What do we agree on?*—leaving the hardest thing for last. During negotiations Biden would have McConnell on speakerphone and put it on mute and yell to a member of his staff: "Can we give that to him?" For Obama, dealing with Congress was an intellectual exercise and not about personal relationships. "It was like being a cook if you don't like food," one Biden aide said.

Obama's disdain for Congress was his Achilles' heel. The Obama West Wing put such a premium on control, on no drama, that it often failed to see the strengths in Biden's improvisational approach. Eventually the West Wing stopped letting him go anywhere near Congress because, Biden aides argue, they were afraid he was going to get something done. "They should have used him a lot more, it was such a gift," said Biden adviser Sarah Bianchi. "I don't think the administration ever understood that all politics is personal." Bianchi said Biden and Obama look at deal-making differently. "For Biden, it's you give me something that makes you want to throw up and I'll give you something that makes me want to throw up and we'll get it done. Obama's view of a legislative compromise is, 'My thing is intellectually correct but if you want to scale it back a little, OK.'"

Biden revered the Senate (in his memoir, *Promises to Keep*, Biden wrote, "Thirty-five years later I still get goose bumps when I come out of Union Station and see the Capitol dome"). He especially enjoyed being president of the Senate and swearing in new senators, taking selfies with them, kissing babies and telling sons and grandsons of senators to "keep the boys away from your sister.'" It's all part of the pomp and ceremony and collegiality he so loved about being in the Senate. It became a running joke, when Biden would come back from the Hill and would be waxing on about a meeting with a senator, Obama would laugh, roll his eyes, and say, "Joe, you just love those guys, don't you?"

Biden had a bad feeling about the healthcare website's rollout. Denis McDonough, then White House chief of staff, came in to brief Biden and his staff on the Healthcare.gov website before it was launched in October 2013. McDonough told the vice president confidently, "It's going to be like Amazon.com, like buying diapers." When he left, Biden turned to a staffer and said, "Something's wrong, right?" It seemed too easy. The staffer agreed and asked, "Are you going to do something?" Biden mentioned his concerns to the president and the president told him, "We have a process." Biden decided he was not going to pick that battle. Healthcare, it turned out, was much more complicated than buying diapers and Biden's instincts were correct. The website, a $630 million online insurance marketplace, met with disastrous results as people were unable to log in and sign up for healthcare coverage, a crisis that helped fuel the backlash to Obama's major domestic policy accomplishment.

Obama relied on Biden for foreign policy advice. Obama also

used Biden's long relationships with foreign leaders and his easy rapport with many of them to help mask some of his own aloofness. Early on in the administration Obama had a lunch with the emir of Kuwait that was stilted and uncomfortable. Then the emir met with Biden, who talked about his grandkids, and at the end the emir was beaming. Biden likes the human give-and-take, while Obama, West Wing staffers say, is probably the least needy person and as a result he does not really understand that most other human beings are essentially needy.

Inside the White House Obama was nicknamed "Mr. Spock," after the human-alien philosopher on *Star Trek*. "This guy is really smart. I've worked with eight presidents and he is by far the smartest," Biden marveled, "just pure gray matter [the part of the brain that processes information]." Sometimes, though, Biden said, Obama needed to be reminded to connect more with his emotions.

Biden was the administration's point person on Iraq from the beginning of Obama's first term. Like other important assignments, he did not find out it was his until Obama announced it during a morning briefing in the Oval Office. "Without saying anything to me, he said, 'Joe will do Iraq. He knows more about it than you guys do. Joe will do it.' At first I thought it was a joke and they did, too," Biden said. "From that point on, everything on Iraq went through me. That surprised me, five weeks in." In 2003 Biden had voted in favor of the Bush administration's U.S.-led invasion of Iraq. The invasion toppled Iraqi president Saddam Hussein's authoritarian regime but it left a vacuum that sparked a massive sectarian civil war. When he was a senator, Biden argued that much of the resulting turmoil was because of ancient hatreds between Sunnis, Shiites, and Kurds, and in a 2006 *New York Times* op-ed he presented a plan with Leslie Gelb, president emeritus of the Council on Foreign Relations, to divide the country along sectarian lines. The plan went nowhere and was sharply criticized at the time for encouraging a splintering of the country along ethnic lines. But some of Biden's aides still argue for it.

Biden became skeptical about Iraq ever finding peace, and he was critical of the notion of imperial America. There are some places, he argued, the United States cannot save. The United States withdrew its combat forces from Iraq in 2011, and since 2014 the Islamic State, also known as ISIL or ISIS, a violent group claiming religious authority over all Muslims, has emerged and seized large parts of Iraq and neighboring Syria. Critics argue that the United States should have forced then–prime minister Nouri al-Maliki, head of a religious Shiite party, out of the government when he began consolidating power and failed to stop ISIS fighters from taking over large parts of the country. Biden wanted to keep Maliki as prime minister because, as his national security adviser Tony Blinken put it, Maliki was the best option. As Maliki's second term unfolded after his reelection in 2010, tensions began building. "The vice president spent endless hours and time trying to prevail on the leadership, starting with the prime minister, to govern . . . in a truly inclusive manner," Blinken said.

Biden's influence on U.S. policy in Iraq is indisputable. He told Christopher Hill, who was U.S. ambassador to Iraq from 2009 to 2010: "My deal with Barack was that I don't have to prevail but I need to be in the room when important decisions are being made." Hill said Hillary Clinton came to the U.S. embassy in Baghdad and "never came again" when he was serving as ambassador. There was criticism inside the White House that Clinton was distancing herself from the quagmire. "From the point of view of a presidential candidate, and everybody knew she was a presidential candidate from the get go, I think she just didn't want to touch Iraq," Hill said.

When Obama came into office he began a review of the war in Afghanistan, which started in 2001, when U.S. and British forces launched airstrikes after the 9/11 terror attacks. By the time Obama took office there were 36,000 U.S. troops in Afghanistan and 32,000 NATO forces. He began to rely on Biden to play the "bad cop" to his "good cop" during talks with generals who pushed for more troops than Biden thought was necessary. The strategy enabled Obama to stay above the fray. Biden was cautious about committing more troops and aides say people would not recognize him in some of these meetings with the defense secretary and other national security leaders—he was exacting, contrarian, and nothing like the affable "Uncle Joe" depicted on the internet. "Obama would start the meeting and it would get to a certain point and Obama would lean back and Joe would lean in," Biden's best friend and Senate successor Ted Kaufman said. "Joe was the devil's advocate." Eventually it got to the point where Obama and Biden would talk in the Oval Office to discuss their joint strategy going into a meeting, and Biden would often have written a note to the president pointing out top concerns to bring up in a meeting, or Obama would say to Biden, "We really have to press on this, you take the lead." That way people, particularly the military brass who were pushing for more troops, did not know which way Obama was leaning. "He's a master at asking questions in a way that you think you know which way he's leaning and then you realize you don't," said Blinken, who was a foreign policy adviser to both Biden and Obama. "Biden was his guy pushing, prodding, challenging. By the last month of the review it was very much a deliberate thing."

But Biden ultimately lost the argument. Before the review was even completed in December 2010, Obama accepted the Pentagon's recommendation to send 17,000 more troops to Afghanistan. (He eventually agreed to send 30,000 additional troops but instituted a timetable for a drawdown in July 2011.) By August 2010, amid escalating violence, the United States had 100,000 troops in Afghanistan. Biden never thought more troops was the answer, and during an August 2015 National Security Council meeting he argued that Afghanistan would revert to chaos no matter how long U.S. troops stayed. "It doesn't matter if we leave tomorrow or 10 years from now," he said. He admitted that he was a "broken record" on this issue. Biden told Hill he was overruled on doubling down in Afghanistan. "He didn't see Afghanistan as something that was going to result in a victory parade down Pennsylvania Avenue."

———

In the late summer of 2013, Obama called his top aides, including Biden, into the Oval Office after a forty-five-minute walk around the South Lawn with his chief of staff Denis McDonough. The two men had been discussing the use of force in the Syrian civil war, which was originally rooted in the 2011 Arab Spring, when revolts toppled Tunisian president Zine El Abidine Ben Ali and Egyptian president Hosni Mubarak. Syrian president Bashar al-Assad responded to protests by killing and imprisoning demonstrators and igniting a civil war that has killed hundreds of thousands of Syrians. A few days earlier more than fourteen hundred Syrians had been killed with sarin gas outside of Damascus and photographs of the evidence revealed that Syrian president Bashar al-Assad's forces had used chemical weapons to kill innocent civilians, including children. Even though the use of such weapons was something Obama had said would cross a so-called red line that would require a U.S. military response, on that walk with McDonough Obama decided to pull back from such a response. Obama told his aides that support from Congress would be necessary to order a military strike in Syria.

Biden was skeptical and says Obama should never have threatened Syria if he did not intend to follow through. Biden, who likes to say "Big nations can't bluff," told Obama, "No more red lines with me, man." One Biden aide described Obama's decision as the starkest disagreement between the vice president and the president in their eight years in office. Biden said he met with 158 members of Congress and spent at least two hours with each of them—either in groups as large as twenty-five or as few as three—to try to convince them, particularly Republicans, to vote for authorization to use force against Assad. The bill to authorize force never received a floor vote in the House or Senate. "He [Obama] did intend to keep the red line, but he really does believe in the separation of powers, and he questioned whether he had the authority to do it," Biden said. "I didn't think it was good to announce [the red line] ahead of time."

At times Biden bristled at the notion that he was being muzzled by the president's West Wing staff, and his aides took great pains to make it look as though he was never being told what to do. But when Israeli prime minister Benjamin Netanyahu was invited by Republican Speaker of the House John Boehner to speak before a joint session of Congress in March 2015, it was clear that Biden would not get to do what he wanted. Biden had known Netanyahu for years and was adamant that, as president of the Senate, he should follow tradition and sit next to the House Speaker behind Netanyahu during the address. But when it became clear that Netanyahu would use the speech to criticize the deal the administration had reached with five other countries and Iran to place limits on Iran's nuclear program, it was decided Biden should not attend the session. What would he do, West Wing staff asked, sit on his hands and not clap when Netanyahu attacked the controversial deal? The White House had not been consulted before then–House Speaker Boehner invited Netanyahu to give the speech and considered that a breach of protocol. Obama would not meet with Netanyahu when he came to address Congress, and no matter how much Biden wanted to, neither would he. It was decided then that when Netanyahu spoke, Biden would be more than fifteen hundred miles away in Guatemala on a two-day visit, meeting with the presidents of El Salvador, Guatemala, and Honduras, and the president of the Inter-American Development Bank.

———

By the end of their eight years together, and especially after Biden leaned on Obama for support during his son's illness, it was clear that the two men had become close. Valerie Jarrett says there was an immediate connection between Jill Biden and the first lady. "Their families clicked instantly. Michelle Obama and Jill Biden, after that first hug, were soul mates. Their children and grandchildren connected." The Obamas even flew to Delaware for the funeral for Biden's mother, Jean, and the president and vice president would go together to the Chevy Chase Community Center, where Obama's daughter, Sasha, and Biden's granddaughter, Maisy, who are close friends, played basketball, and sometimes the president would coach. Biden joked, "I warned the president when he got me he got the whole family."

During their weekly lunches Obama and Biden talked about Beau's illness. Biden felt a responsibility to the president to make sure he understood what was going on. It helped to strengthen their relationship in an unprecedented way. "He's the only person outside my family I ever told about Beau," Biden said, his eyes filling with tears. "I'd sit there and talk about Beau at lunch, and I stopped talking about it as much because he would start to cry."

After Beau's death, about ten staffers gathered at the Observatory to help the vice president and his wife plan Beau's funeral. Biden told the group that during one of their weekly lunches, Obama had offered to pay off his mortgage when Biden told him that he and Jill were considering selling their house to help support Beau's wife, Hallie, and their two children. "It seemed cathartic for him to talk so no one interrupted him. Not that you would interrupt him anyway," said one Biden aide, who did not want to be identified because of how intimate that moment was. Biden later recalled the conversation he had with Obama in a television interview: "But I worked it out," he said he told Obama at one of their lunches. "Jill and I will sell the house and be in good shape."

"Don't sell that house. Promise me you won't sell the house," Obama pleaded with Biden. "I'll give you the money. Whatever you need, I'll give you the money. Don't, Joe—promise me. Promise me." The house is one of the only things Biden owns, so

Obama's offer to help him was especially meaningful. (In the end, Biden said, he did not need to take Obama up on it.)

At Beau's funeral Obama said, "Michelle and I and Sasha and Malia, we've become part of the Biden clan. We're honorary members now. And the Biden family rule applies. We're always here for you, we always will be—my word as a Biden." In Biden's big Irish Catholic family, Obama found what he never had growing up without a father at home. "Family has been central for us—that's our baseline," Obama said of Biden. "We both feel freer to do what we think is right because if it doesn't work out, our families will still love us."

President Obama's decision to include Biden's Cancer Moonshot, a national effort to end the disease by increasing research funding in his final State of the Union Address was a sign of their strong partnership. Biden's chief of staff at the time, Steve Ricchetti, said it meant a lot to Biden because the effort "was a defining part of what he intended to do, not just in the remainder of his time in the vice presidency, but for the rest of his life." But the most dramatic example of their friendship came days before they left office, when Obama surprised Biden with the Medal of Freedom, the nation's highest civilian honor, created by President John F. Kennedy in 1963. Biden is the third vice president to receive the medal—the others are Nelson Rockefeller and Hubert Humphrey, who received his posthumously.

Obama had been planning the surprise for months; he and his staff had considered including it as part of a bigger ceremony at the White House in November 2016, and then again at a bill-signing event in December, but ultimately it was decided that it should be a stand-alone event. Only Biden's wife, Jill, and Ricchetti knew in advance. It was treated like a national security secret. The invitation billed the event as a "toast to the Bidens," and the event programs printed with the "Medal of Freedom" title and citation were closely guarded and transported from the printing office to the White House with an escort. Two versions of the president's remarks were circulated and put into different briefing books for the president and senior staff, with only a very small group given the books that included Obama's Medal of Freedom remarks.

"This is the kind of family that built this country," Obama said at the White House ceremony honoring his vice president. "That's why my family is honored to call ourselves honorary Bidens." The medal came with an extra level of significance because it was the only time in Obama's presidency that he awarded it with a level of distinction, a designation most recently given to Pope John Paul II, President Ronald Reagan, and General Colin Powell. "He was a brother in arms and a valued counselor and a troubleshooter who the president could turn to," Axelrod said of Biden. Both the president and the vice president had tears in their eyes at the medal ceremony. "You kind of fall in love with him [Biden]," an aide said, "and I think Obama did."

XIII

Man on a Wire:
Mike Pence's Tightrope Act

When we were first talking about whether he should be
Trump's running mate, he said Trump reminded him
of our dad's personality. That surprised me.

—MIKE PENCE'S OLDER BROTHER, GREG

If you look at a corporation, the number two person is the one most
people report to. And you ignore him or her, or are not deferential to
him or her, at your peril . . . Not the case with Trump and Pence.

—A PERSON WITH INTIMATE KNOWLEDGE OF THE RELATIONSHIP
BETWEEN PRESIDENT TRUMP AND VICE PRESIDENT PENCE

"He's a huge value added for us. We all know him. He has, I think
we'll all stipulate, a very different kind of personality from the
president and he's in the middle of everything and it's been great."

—SENATE MAJORITY LEADER MITCH MCCONNELL

A re you married?" a top White House aide asked rhetorically. "I spent the first couple years of my marriage thinking I needed to change my wife. After a while, friends
said, 'You fell in love with her and you knew she had all those flaws,
all those characteristics, and it's not your job to change her.'" That,

the person says, is how Mike Pence feels about Donald Trump. "These are old guys. They're not going to change."

Vice presidents are usually the attack dogs during campaign season. It is a role they tend to relish, from Nixon to Cheney to Biden. But Trump and Pence reversed these roles, with Pence serving as the voice of restraint and Trump the voice of outrage. But in this case the outrage is extreme. One former vice president, who did not want to go on the record, pointed out the unusual dynamic between the bombastic Trump and his understated vice president: "What would be unacceptable would be if the roles were reversed and Pence was the guy sending tweets all the time and going after people. The role of being the attack dog when you're vice president, it's a useful role to play; somehow this is reversed, at least for now." The suggestion that Pence is a "peer" of Trump's is disingenuous, said a member of Trump's Cabinet. "All vice presidents have a secondary role and the role is to support to the very best of their ability the president they serve and to reinforce the president's positions."

For his first trip abroad as vice president, Pence was given a particularly challenging assignment. He was dispatched to attend an international security conference in Munich, Germany, to reassure nervous NATO allies that the United States had no plans to abandon them, even as his boss railed against members of the powerful alliance and accused them of not paying their fair share of the cost of mutual defense. Pence knew he had to choose his words very, very carefully. A draft of his remarks was run by President Trump's national security team and it was approved. But Pence was still worried. *Would any of it upset his boss?* He insisted on running the speech by Trump himself. After they landed in Munich and had made their way to the hotel, Pence's aides got the president on the phone at 1:00 a.m. Munich time. Pence read through the speech line by line, and they edited it together. Pence's nickname inside the West Wing is "on-message Mike" for a reason.

He is unfailingly polite and deferential and has made an art form out of avoiding the spotlight, knowing how angry Trump gets when others take credit for his success. Even though his press staffers have a Google alert set up so that they see whenever their boss is mentioned, Pence will often read stories about himself before anyone else sees them. He does not want to be the center of attention, at least not right now, but, like his boss, he is an avid media consumer. Pence is uniquely suited to working for a boss like Trump, who demands absolute loyalty. Early on election night, when things were not going well for the Trump/Pence ticket, Pence texted a "Dewey Defeats Truman" cover to some of his pessimistic friends. When he stands beside the president he has a look of pure devotion and, with that, subservience. "Mike wouldn't want it printed in a book that he's a steady hand because that might be viewed by the president in a way it shouldn't be," said a member of Trump's Cabinet. "He might think it would make it look like he doesn't know what he's doing so he has to turn to Mike. But personally, I think Mike *is* a steady hand and a calming force."

Before he left for a week-long trip to South and Central America last year there was a sudden ramping up of tension with Venezuela. Days before the trip, Trump said he was considering a "possible military option" in the country to stem the turmoil there. "President Trump is a leader who says what he means and means what he says," Pence said at a joint news conference with Colombian president Juan Manuel Santos. But privately Pence made it clear that that is not always the case. U.S. allies in South America were practically begging for reassurance from Pence that Trump was not serious about military intervention. "What do you think an oil embargo is?" Pence told them, seeking to calm their nerves. "You don't do a naval blockade with commercial ships. That's what he *could* mean by military intervention." Trump did not tell Pence to do this, no one inside the West Wing seemed to care particularly about allies panicking, but Pence views his role as vice president to act in part as Trump's translator. At the White House, Pence summarizes meetings for Trump. According to a senior Trump official,

Pence will cut in after an hour of debate and say, "Mr. President, this is how I see where the group thinks we ought to go."

Pence canceled a previously scheduled interview for this book after he flew to Texas in August 2017 in the aftermath of Hurricane Harvey. When Pence went to visit victims two days after Trump's visit, he was emotional and connected with their grief in a way that the president had not. Headlines screamed "PENCE SHOWS TRUMP HOW TO SWEAT IT OUT WITH TEXAS VICTIMS" and "MIKE PENCE'S 2020 RUN GOT OFF TO A GREAT START IN TEXAS." He had earlier released a statement calling an August 2017 *New York Times* story suggesting he was positioning himself for 2020 "disgraceful and offensive." The statement had an audience of one: Donald Trump. He tells aides, "stay in your lane," meaning *do not step on the president's message.* When Pence helped push the Republican-controlled House to pass its own version of a healthcare bill in the summer of 2017 (it later failed to pass in the Senate), Pence aides asked lawmakers and outside groups to downplay his role. When Trump ordered him to leave an October 2017 NFL game because some players took a knee to protest police brutality and racial injustice during the playing of the national anthem, Pence and his wife, Karen, did as they'd been told by the president: they walked out. "His job is defined by one person, and the most important thing is to protect that relationship," said Pence's first vice presidential chief of staff Josh Pitcock, who worked for Pence for twelve years and left after six months on the job. Pitcock was replaced by political operative Nick Ayers.

Ayers works out of a glorified closet in the West Wing. The hallways of the West Wing are decorated with heavy mahogany furniture, Asian lamps, and large photos, known as "jumbos," of Trump taken by the White House photographer (which is a long-standing tradition). Ayers is unfailingly loyal to both Pence and the president he serves—nothing gets him angrier than stories that make the case that Pence is plotting his own political future. When Trump and Pence have lunch they are accompanied by Trump's chief of staff John Kelly and by Ayers, who act as chaperones—it is a highly unusual setup. Those weekly lunches between presidents and their vice presidents are so important to modern vice presidents precisely because they offer precious time when they can be alone and talk candidly with the president. But Trump and Pence need the conversation to be steered by Kelly and Ayers so that it does not get off track, and sometimes Ayers will interpret remarks Trump makes for Pence later on after lunch.

Trump never holds back on taking digs, and Pence is no exception. One Pence aide recounted a Situation Room meeting: "The president made a comment, 'I know Mike wouldn't say it this way,' and he said it the way Donald Trump says things, and then he said, 'But Mike, how would you say it?' Pence rephrased the same statement in a more politically correct way and Trump turned to the larger group and said, 'See, *that's* what I love about him.'" According to a profile of the vice president in the *New Yorker,* during a private meeting with an unnamed legal scholar, when the topic of conversation turned to gay rights Trump motioned to Pence and allegedly joked, "Don't ask that guy—he wants to hang them all!" Trump was once fascinated by the Pences' piety and now mocks it in the White House asking visitors to the West Wing, "Did Mike make you pray?" According to former chief Trump strategist Steve Bannon, it's Trump's way of letting "Pence know who's boss." A good friend of the Pences' said she would not be surprised if Trump nicknamed his vice president "the deacon." The teasing, not surprisingly, is not mutual. Pence never jokes back and every time he leaves the Oval Office he thanks the president for his time. No matter how difficult their conversation was or how inappropriate the joke.

A Republican lobbyist described attending a meeting in the White House's Roosevelt Room after Trump nominated Neil Gorsuch to a seat on the Supreme Court. Pence walked into the room a little late and stood in the back as the president spoke, looking unsure about whether he should go to his seat. When Trump asked him where he'd been, Pence said, "I took your nominee up to the Hill and here I am." The president thanked him and said he could not have won the election without Pence and referenced a couple of specific rallies where Pence had done well. Pence then sheepishly took his seat. Pence and Trump talk at least once a day, said Pence's first White House spokesman, Marc Lotter, who resigned after less than a year on the job. Lotter said the two have grown close, but he could not offer a single specific example of their personal or professional camaraderie.

Even though the president's tweets upset some inside the White House, Pence does not weigh in on them, though he might be more supportive of them than most people suspect. When asked what the vice president's reaction was to a particularly vicious tweet about Senate Majority Leader Mitch McConnell, one aide said Pence's thinking is: *If you do what we say, then there will be no more tweets.* Joe Biden, who has been offering Pence advice, said he does not think Pence necessarily disagrees much with the president. "I think it's the real him not taking on the president . . . when I saw Mike and his wife walk out of the NFL game, it may have been planned, but Mike probably thought they [the players] shouldn't have been kneeling."

Pence seems resigned to the president's behavior and a willing accomplice to it. "I do know that many people who are very close to the president have asked him back through the campaign to today to, if not refrain, at least be more disciplined," said Bill Smith, who was Pence's chief of staff when he was governor of Indiana and who remains extremely close to the vice president. "But the president will do what he will do."

Pence also reaches out to Dick Cheney for advice, but Pence aides say their boss is more closely modeling his vice presidency on that of George H. W. Bush. Pence's ability to tame the president, or to define himself amid the chaotic administration, has fallen short. Trump did not give Pence a specific portfolio, they did not "divide up the world" as one Pence aide put it. Instead, the vice president's portfolio is far-reaching yet amorphous and unclear. "Everybody appreciates that he [Pence] has the president's ear," said Brendan Buck, the top communications aide to Speaker of the House Paul Ryan, "and that he's an enduring and respected part of his operation, an operation that is somewhat in flux." When Trump attacks Republican allies on the Hill, including Ryan and McConnell, Pence's role as intermediary is made more complicated.

"When we were first talking about whether he should be Trump's running mate, he said Trump reminded him of our dad's personality," recalled Pence's older brother Greg. "That surprised me." But, Greg explained, after he thought about it for a while it began to make sense: "You always knew where you stood with my dad, he was very direct. He just said what he thought, period." Pence's father, Edward Pence Jr., was a decorated Korean War veteran whom Pence idolized—he keeps his father's Bronze Star in his office. But he was something of a bully and was a strict disciplinarian, who used a belt to hit his children when they misbehaved. At dinner, the six Pence children were not allowed to speak, and if they forgot to stand when an adult entered the room, he would grab them and force them up onto their feet.

Trump is thirteen years older than Pence and, as it was with Pence's father, Trump does most of the talking during their private meetings. Pence and Trump have a relationship more like the domination-subordination that characterized the alliance between Lyndon Johnson and Hubert Humphrey than the friendship that evolved between Obama and Biden, or the professional partnership during the Bush-Cheney years. Pence even occasionally gives lessons to other members of the administration on how best to handle Trump. When then–secretary of state Rex Tillerson was on the brink of resigning in the summer of 2017 because of frequent clashes with Trump, it was Pence, according to an aide, who coached him on how to handle his relationship with the president: air concerns in private, he said, and be respectful in meetings. So Tillerson, the former CEO of ExxonMobil, who was fourth in line to the presidency, was getting a pep talk from the man who is first in line and who has had lots of experience dealing with domineering personalities.

Even top aides admit that Pence is loyal to a fault, sometimes standing by and defending Trump even when it jeopardizes his own reputation. Typically, vice presidents do not sit in the audience at presidential news conferences, but Pence is a perennial presence in the front row. Pence believes in what he calls "servant leadership" and tells his staff to follow the three keys to leading successfully: humility, self-control, and orientation to authority. Indeed, it is his unwavering loyalty to the president he serves that has been a main point of criticism. As a member of Congress from 2001 to 2013, Pence established himself as a solid conservative who backed free-trade deals—in 2001, he praised the North American Free Trade Agreement, a deal that Trump has vowed to "terminate" if it cannot be renegotiated. Pence was not in Congress when NAFTA was passed in 1993, but he did vote for the Central American Free Trade Agreement in 2005 and he also voted for free trade agreements with Colombia, Panama, and South Korea in 2011. As with his speech to NATO allies, he faces a particularly difficult assignment: how to stay true to himself and his values while being loyal to Trump. He saves most of his advice and counsel for one-on-one meetings with Trump in the Oval Office. His aides work hard to keep Pence out of the high-stakes drama that unfolds daily in the White House. There are too many leakers in the White House, they say, and if the president makes a different decision than the one Pence recommends, and that gets out in the media, it would make Pence look weak and undermine his influence. "He does not accept palace intrigue," said Lotter, before he left the White House. "It's cultural, and it stops those types of distractions from consuming our office. No one in our office looks around and thinks rumors are coming from the person sitting next to them."

At a 2017 commencement address for graduates of Pennsylvania's Grove City College, a Christian school, Pence declared, "Servant leadership, not selfish ambition, must be the animating force of the career that lies before you." He continued: "Don't fear criticism. Have the humility to listen to it. Learn from it. And most importantly, push through it. Persistence is the key."

————

Michael Richard Pence was born in 1959 in Columbus, Indiana. He was one of six children raised in a devout Irish Catholic family. His father ran a chain of convenience stores and his mother was a homemaker. Pence was an altar boy and went to parochial school. "Our life revolved around the church," his brother Greg says. He developed a love for public speaking early on and did not hide his political ambition. His high school speech coach, Debbie Shoultz, said, "Even when he was a senior, he talked to his classmates about one day being president." His grandfather, who immigrated to the United States from Ireland in 1923, taught Pence to revere legendary Democratic presidents such as FDR and JFK. Pence's first vote in a presidential election was for Democrat Jimmy Carter, who ran against Ronald Reagan in 1980. Pence voted for Carter, he explained, because Carter "was a good Christian." In a now-ironic follow-up he added, "Beyond that, there was a sense of 'Why would you elect a movie star?'"

He went to Hanover College, a small liberal arts school in Indiana, with an interest in broadcast journalism, but the most notable part of his time there was his spiritual awakening. "Mike became deeply religious during that time," his mother, Nancy, recalled. "At one point he was seriously considering entering the priesthood." But he felt a void in his life and strained against the ceremony-laden Catholic Church. "I began to meet young men and women who talked about having a personal relationship with Jesus Christ," he said. "That had not been a part of my experience." He admired the gold cross on the neck of his fraternity "big brother," who told him, "Remember, Mike, you have got to wear it in your heart before you wear it around your neck." At a Christian music festival in Kentucky in the spring of 1978, Pence made the decision to leave the Catholic Church and become a born-again Christian. "I gave my life to Jesus Christ," he recalled years later about that pivotal concert, "and that's changed everything." It was the beginning of a difficult transition, though, one that would take years and cause strain in his family.

After graduating from Hanover, Pence entered law school at Indiana University, where he met Karen Sue Batten's sister, who would connect them. Even though he had converted to evangelical Christianity in college, the Pences were married in a Catholic church in 1985. He referred to himself as an "evangelical Catholic" until the mid-1990s. Pence received his law degree in 1986 and went into private practice, around the same time he became a Reagan Republican. A year later, in 1987, he made a decision that surprised even his own family: he ran for a seat in the U.S. House of Representatives—he was twenty-nine. Pence's father was so against it that he even asked one of Pence's brothers to help convince him not to run. Eventually, though, Pence won his father's support. He perfected his aw-shucks, folksy modesty during that early campaign and rode a single-speed bicycle around his district in shorts and sneakers striking up conversations with anyone he could find. He won the Republican Party's nomination but lost the general election by a slim margin to Democratic challenger Phil Sharp. He ran again in 1990, and it was a vicious campaign on both sides: the Sharp campaign seized on Pence's use of campaign funds to pay for living expenses, and Pence upped the ante by running a television ad showing a man dressed as a cartoonish version of an Arab in a robe and headdress and standing in front of a fake backdrop meant to look like the desert. The actor praised Sharp, who was chairman of a subcommittee dealing with energy issues, for being the "best friend" of Arab oil-producing nations. The ad was roundly criticized. The *Indianapolis News* said the actor had used "perhaps the worst Omar Sharif impersonation ever recorded" and called it "insulting and derogatory . . . appealing to racist sentiments."

Pence lost to Sharp again, this time by nineteen points, and a year after his defeat he wrote an essay, "Confessions of a Negative Campaigner," published in the *Indiana Policy Review.* Pence began the essay with a passage from the Bible: *It is a trustworthy statement, deserving of full acceptance, that Christ Jesus came to save sinners, among whom I am foremost of all.*—1 Timothy 1:15. "Negative campaigning is wrong," he wrote. "It would be ludicrous to argue that negative campaigning is wrong merely because it is 'unfair,' or because it works better for one side than the other, or because it breaks some tactical rule. The wrongness is not of rule violated but of opportunity lost. It is wrong, quite simply, because he or she could have brought critical issues before the citizenry." He still regrets going against his own moral principles in the 1990 campaign. But, Sharp said in a recent interview, he cannot recall ever receiving a personal apology from Pence. "He's what I call 'Midwest nice.' You don't treat people meanly or nastily to their face. But that doesn't mean you aren't mean and nasty. It's a veneer of niceness."

After losing both congressional races, Pence turned his attention to radio, where he became a talk-show host and built a name brand in Indiana. He would also host a Sunday-morning political television show. One of his earliest radio endeavors was at a small local FM station in Rushville, Indiana, where he cohosted a once-a-week dinnertime political radio show called *Washington Update.* At the end of the broadcast Pence would sign off with: "Good night, Rush County, and good night, Washington. Are you listening?" On air, he was known for his polite demeanor and began calling himself "Rush Limbaugh on decaf." "We'll just have to agree to disagree," he would say soothingly during political arguments. Pence eventually left the small station and worked his way up to Indianapolis-based WIBC, a bigger network with affiliates that spanned the entire state. "He could not be a more *directable* talent," says Kent Sterling, who was an assistant program director for *The Mike Pence Show.* When Pence first started out in radio, Sterling asked him to try not to be too long-winded. "He nodded when I asked him and I wondered if he got it. The next day he was spot on. He was perfect and he never deviated." That same determination, ambition, and attention to detail has made him a successful politician. "He is perfect for Donald Trump," Sterling reflected. "As this administration operates in this chaotic space, he is not going to be chaotic. Mike always understands what his role is." He is a hard worker, said Greg Garrison, who has known Pence for more than twenty years and worked with him in radio. "That's how he sold that show. He went all heartland on them." Pence eventually expanded the show to eighteen markets.

Steve Simpson, a former Pence colleague and Indiana news anchor, said he thought Pence made a brilliant career move by using his radio show to rehabilitate his image. "He ran such a dirty awful campaign and he knew it," Simpson said. "The radio show was a mea culpa." Clips from the show that have been salvaged reveal Pence's prescient early criticism of the mainstream media and adoption of the anti-Washington sentiment that helped get Donald Trump elected. He could be sharp-tongued from time to time. When news broke in 1996 that a sexual harassment lawsuit filed by Paula Jones against President Clinton would be moving forward, Pence said it would "make the O.J. Simpson trial look like traffic court."

In an August 1997 column posted to his radio show's website, Pence supported a House GOP effort to remove Newt Gingrich as Speaker of the House for ethics violations. "Whether Republicans want to admit it or not, House Speaker Newt Gingrich has been knocked off his horse and been wounded badly, maybe mortally," Pence wrote of the man who would become his main competition for the vice presidency. "If the G.O.P. is to find its way back in time for the next election, it is time for new leadership, either a new Speaker or a revived Speaker. I'll take either one." His anti-establishment streak went off the rails when, in a 2000 op-ed, he made the nonsensical claim that cigarettes are not lethal. "Time for a quick reality check," Pence wrote. "Despite the hysteria from the political class and the media, smoking doesn't kill. In fact, two out of every three smokers does not die from a smoking related illness and nine out of ten smokers do not contract lung cancer." In 2000, Pence received more than $10,000 in contributions from the political action committees of major tobacco companies, including Philip Morris and R.J. Reynolds. Over his twelve years in Congress, he received $39,000 from R.J. Reynolds, the maker of Newport and Camel cigarettes. When he ran for governor in 2012 and 2016 (he dropped out of the race to become Trump's running mate) he received more than $70,000 from the tobacco industry.

His radio show was clearly a stepping-stone, a way of keeping his name before the public. No one at WIBC, said former colleague Steve Simpson, thought he would be there for long. "His political ambitions were never more than an inch away from that microphone." One small sign of this, Simpson said, was that Pence "didn't dress like the rest of us." Radio talk-show hosts typically dress very casually, but not Pence, who wore button-down shirts and carried a briefcase to work. "He was looking down the field when he decided to go on the radio." Like most of Pence's friends from Indiana, Simpson thinks Pence is "perfect" for Trump: even-keeled and patiently waiting for his chance to be president. "He is calmly and methodically doing what he's doing. I keep thinking to myself, *Is this like the radio show? Is this marking time as vice president to the next obvious step?*"

———

Now well known in Indiana, Pence was finally elected to Congress in 2000, where he would serve for six terms. His platform included a vow to oppose "any effort to recognize homosexuals as a discrete and insular minority entitled to the protection of anti-discrimination laws." He won by twelve points. "We felt that 2000 was an opportunity to do it the right way," Karen said. Never far from the medium that got him elected, Pence installed a special digital phone system in his congressional office so that when he called into radio shows across the country—which he did often—the sound quality was so good that it seemed like he was in the studio with the host of the show.

In the House, Pence became an early voice of the Tea Party movement. He likes to say of himself, "I'm a Christian, a conservative, and a Republican, in that order." He was a dedicated representative of the most conservative Republicans in the House and started climbing the ranks as deputy whip and then as leader of the Republican Study Committee, an influential group of conservative House members. After Republicans lost control of the House in the midterm elections in 2006, he challenged John Boehner for the position of minority leader but lost by an embarrassing vote of 168 to 27. In 2008 he went after his party's third most powerful position in the House, that of Republican conference chairman, and won. His job was to reshape the Republican party's messaging after widespread losses. On the wall of the conference's main meeting room he hung his personal credo:

MIKE PENCE'S CONFERENCE RULES:

- *Glorify God*
- *Have a Servant's Attitude*
- *Promote Ideas*
- *Promote House Republicans*
- *Have Fun*

"He is a relationship guy and finds great pleasure in having relationships with people and investing in people," said Republican congresswoman Marsha Blackburn of Tennessee, who has known Pence since 2002. "The vice president would always use a phrase with us in the RSC [Republican Study Committee] and then with the conference: iron sharpens iron, the biblical reference. And you come to that by having robust debate and conversation among yourselves and bringing your A game."

Unlike the families of most members of Congress, when Pence first took his seat in the House his wife and three children moved with him to a Washington suburb in Virginia instead of staying home in their congressional district. It was Karen's first time living outside of Indiana. Karen said she asked Dan and Marilyn Quayle for their advice, and they told her that if Mike went to Washington alone and she stayed with their three children in Indiana, whenever he came to a soccer game, constituents would be angling to speak with him. "In D.C., you're no big deal," Karen said. "That appealed to us. We didn't want our kids to grow up thinking they were celebrities. Our priority was our family." Senator Roy Blunt, a Republican from Missouri, remembers dropping Pence off at his house after work and seeing his son playing out in the street with other kids in the neighborhood. Pence, he said, is a family man. "He always said he grew up in a house with a cornfield in the back-yard. He never suggested it was *their* cornfield."

———

After twelve years in Congress, and after losing his leadership bid, Pence thought he needed executive experience as governor in order to one day run for president. He was elected governor of Indiana with less than 50 percent of the vote in 2012. While he used to enjoy a couple of beers with friends back when he was president of the Phi Gamma Delta fraternity in college and in law school and when he was a local radio host, that stopped once he became a member of Congress. His friend and former chief of staff when he was Indiana's governor, Jim Atterholt, said Pence's decision to stop drinking is because he wants "to have his wits about him" in case something happens. Pence only grew more disciplined as governor: "There was no drinking period in the governor's office," Atterholt said. And having a meal alone with a woman or attending events where alcohol is served when his wife is not there has been strictly off limits since he was elected to public office. "He was mindful of appearances," Atterholt says. Friends say his strict rule against dining alone with women was largely at his wife's request, and that the decision to follow what is known as the Billy Graham rule was made out of respect for her and their marriage. Christy Denault, who was Pence's communications director when he was governor, said she completely understands his rationale. Pence wants to protect himself from any suggestion that he is having an affair. "Pence with Mystery Woman" was the headline he and his staff were worried about. Rumors of infidelity, they reasoned, could derail his career. "It's not about morality," said his friend Bill Smith, "it's about his relationship with his wife." Denault said that as governor Pence "did not have closed door meetings with a woman unless there was a third party present." Occasionally he would go into her glass-walled office to have a private conversation.

As governor, Pence kept his Bible open on his desk every day. He still often makes references to scripture; friends say he is not passionate about any one particular issue, but he is passionate about his Christianity. As governor he insisted that meetings be opened with a prayer—he would go around the table and randomly pick a staff member and ask them to lead the prayer. One day a staffer who was not outwardly religious joked, "How come you never pick me to pray?" For the next five meetings Pence called on him. "He'd talk about his Bible study that morning and how that informed him about an issue that he was having to decide on later that day," Atterholt recalled. Pence would cite specific passages and say how a reading could be applied to a policy decision. "His Bible study informed him throughout his day, every day. He has tremendous discipline." As vice president he hosts a lunchtime Bible study, led by an evangelical pastor, with members of the Cabinet at the Eisenhower Executive Office Building, where his ceremonial office is located. Bill Smith said that friends often ask him how Pence finds the time now that he's vice president, and Smith tells them, "If there's a meeting he has to say no to, OK, he has to say no to many meetings." Trump begins most Cabinet meetings with a prayer and usually asks Pence to lead it.

Pence often tells younger staffers: "Don't be a slave to the office. If you spend 20 hours a day in here that says to me that you're not going to be investing yourself in your families." On the rare occasion when he does raise his voice, he usually apologizes for it, and when he is in a bad mood he laughs it off and calls it a "bad hair day." Scott Pelath, who was Democratic minority leader in the Indiana House, said Pence has a sense of humor that he tries to hide in formal settings. One time when Pelath was at the governor's mansion for dinner, Pence told him that a guest had to bow out at the last minute. As a joke, Pelath said, "I always wanted to meet Jim Nabors [referring to the late gay actor who played Gomer Pyle on *The Andy Griffith Show* and who used to sing "Back Home Again in Indiana" at the Indianapolis 500]." An openly gay actor would have been an unexpected guest at the Pences'. Pence wanted to laugh, Pelath said. He knew it was intended to be funny, but he just could not do it. "He has such a rigid sense of what's appropriate in his worldview."

Pence, Pelath says, is a man who values the trappings of office, the pomp and the spectacle. He said when he went to Pence's office the chairs were always carefully aligned; the photographer was always in the room to take pictures and record every important event. This fits into Pence's calculated approach to political life. "To Mike Pence, it doesn't matter what you're saying, as long as you're standing in the right spot when you're saying it," Pelath said. "I've never questioned Mike Pence's convictions with respect to his faith," he continued. "He's gotten exactly what he's wanted in his life of faith. I'm worried that it's made him even more resolute in his belief that he's been sent on a holy mission. And that is dangerous for the country. It's untenable."

Pence's job approval rating in Indiana dipped below 50 percent shortly before he was picked as Trump's running mate in 2016. Pence has regretted how he handled some issues in his state, including a plan to launch a state-run news agency that prompted the headline "PRAVDA ON THE PLAINS." The single biggest controversy was his signing of the Religious Freedom Restoration Act, or RFRA. It became a defining part of his tenure as governor of Indiana. When Pence initially signed the bill in 2015, it sparked national opposition from business leaders and the LGBTQ community, which said the measure would allow businesses to discriminate against gay people. Pence never anticipated the backlash.

At the private bill signing in his office Pence was surrounded by monks and nuns in habits and highly controversial figures from Indiana's most conservative groups, including Indiana Family Institute's Curt Smith, who has said homosexuality is outlawed in the Bible. "I don't know if you want to call it a defect, but that element of his personality is what caused the entire debacle because an ordinary Indiana governor would take a bill like RFRA, have a sense it was controversial, wait till the legislature went home, and quietly sign it amid a stack of fifty other bills," said Pelath. "But he couldn't resist. He had to have the photo." Indiana state representative Ed DeLaney, a Democrat, was accused of Photoshopping the photo when it was circulated because his colleagues thought it was so outrageous.

As with Trump, it is not always clear how Pence makes decisions. Most of his staff was with him late one night before a disastrous interview on ABC's *This Week with George Stephanopoulos* in which he tried to defend RFRA. He and his staff had reached a consensus the night before that he would not be doing the interview. But Pence called aides early the next morning to tell them he had changed his mind and would go on the show. (Advisers speculate that Karen Pence changed his mind.) When Stephanopoulos pressed him on whether the law allowed discrimination against gays and lesbians, he could not answer.

As businesses threatened to leave the state, the law was eventually rewritten, prompting a backlash from some religious conservatives who accused Pence of flip-flopping on the issue. But conservatives would have trouble questioning Pence's bona fides. He supported "personhood" legislation that would ban abortion under all circumstances, including rape and incest, unless a woman's life was at risk. A year later, in 2016, he signed a bill that made Indiana the second state to ban abortions sought because the fetus had a disability, and that required the remains of aborted or miscarried fetuses to be interred or cremated. It was blocked from going into effect by a federal judge who ruled that it was likely unconstitutional.

––––––––

Pence's critics are quick to point out his obvious ambition. "Thematically everything is ideologically based but individual decisions at any given time are weighed very specifically against how is this going to get me closer to being president of the United States," said Bill Oesterle, who managed Republican Mitch Daniels's campaign for governor in 2004 and supported Pence's gubernatorial run in 2012.

Before the 2012 presidential campaign, a group of influential conservatives who were dissatisfied with likely Republican nominee Mitt Romney met with Pence in his Capitol Hill office when he was still a member of Congress and had not yet decided to run for governor. In that meeting, according to former Indiana congressman David McIntosh, Pence "acknowledged he had always wanted to be president." And for a couple of months he considered their efforts to recruit him. But he ultimately decided he needed executive experience before running for president. He was concerned that first-term Alaska governor Sarah Palin had lacked sufficient executive experience when she was John McCain's running mate in 2008 so he decided to run for governor instead.

In Pence's first year as vice president it was unclear whether he saw himself facing a dilemma as the right-hand man to such a combative and deeply divisive president. Based on dozens of conversations with his friends and detractors, he saw Trump's offer as a lifeline and not as a moral dilemma. "No unblinkered observer still can cling to the hope that Pence has the inclination, never mind the capacity, to restrain, never mind educate, the man who elevated him to his current glory," conservative columnist and unabashed Trump critic George Will wrote in an op-ed in the *Washington Post*. "Pence is a reminder that no one can have sustained transactions with Trump without becoming too soiled for subsequent scrubbing."

But it was a deal Pence was all too willing to make. Shortly after Trump's election, Pence was named head of the transition team. He asked his staff for every detail they could provide on so-called continuity of government, the procedures that dictate how the government would operate in case of a catastrophe that took out the president. *My job*, Pence often tells aides, *is to be prepared*. His most obvious influence on the Trump administration was in staffing top positions in the government. He personally picked and advocated for former Indiana senator Dan Coats to become director of national intelligence (Coats says he delayed retirement for the job and would not be working in the White House if it were not for Mike Pence); for former Georgia congressman Tom Price to be Health and Human Services secretary (Price texted Pence to tell him he wanted the job, but he was fired in 2017 because of his use of private jets at taxpayer expense); for Kansas congressman Mike Pompeo to head the CIA (Trump later picked Pompeo to replace Rex Tillerson as secretary of state); for longtime friend and former chief of staff at the House Republican Conference Marc Short to be Trump's director of legislative affairs; and for Indiana consultant Seema Verma to head up the Centers for Medicare and Medicaid Services. Pence has friends across the federal government.

Lobbyists and lawmakers know Pence's value and influence and they will wait for hours to have a ten-minute talk with the vice president, knowing that he might get called into a last-minute meeting with the president. They bide their time on red couches in the White House lobby or they sit on the gold and blue chairs in Pence's West Wing suite.

Pence's approach to the vice presidency mirrors his advice to U.S. Naval Academy graduates. In his May 2017 commencement address at the academy he talked about "orientation to authority" as a key part of leadership. "Follow the chain of command without exception. Submit yourselves, as the saying goes, to the authorities that have been placed above you," he said. "Trust your superiors, trust your orders, and you'll serve and lead well." And revere your superiors: Pence often talks about Trump's "broad shoulders" and praises his "strength."

Pence is keeping his nose down and staying out of Washington as much as he can. Much like Gerald Ford, who was vice president during a tumultuous time, Pence has packed his schedule with foreign travel and fund-raisers for congressional candidates. As former FBI director James Comey testified in Congress—charging that the president lied and tried to stop investigations into former Trump national security adviser Michael Flynn and his business dealings with foreign countries during the presidential campaign—Pence was telling a group of governors and state officials about the administration's commitment to building new infrastructure. "Folks," Pence proclaimed, "it's already been a banner week for infrastructure." Pence has hired a personal attorney to represent him in inquiries relating to the federal investigation into possible ties between the Trump campaign and Russia. Pence's older brother Greg said, "My brother does not believe there is any truth to the Russia allegations." The two have discussed it and, he added, speaking candidly for himself, "If it's true, then we need to find out." But one former high-ranking Republican official said Pence's absence from a meeting with the then–Russian ambassador to the United States Sergey Kislyak—a meeting that Trump invited Dina Powell, who was his deputy national security adviser, to attend—had a flip side: "The question is, I hope he was either on the Hill or out of town. Otherwise, he should have been there." If he is a top presidential adviser, he should have attended the meeting.

Pence's unprecedented decision to form his own political action committee within the administration's first six months suggests he is preoccupied with his own future. He is not yet sixty, and aides and the party elite believe he could one day be president. His PAC, the Great America Committee, can support candidates, including himself and Trump, in 2018 and 2020. Pence raised an astonishing $1 million for his leadership PAC during a single evening cocktail party in July 2017. Before the July event, Pence's PAC disclosed raising $540,071 in the first six weeks of its creation in May 2017. Dozens of corporate PACs, including AT&T and Boeing, have given to Pence's group, hoping that he will grant them access to the West Wing. The process in the West Wing of getting information to the president is chaotic, and Pence and his team are not viewed as a competing power center, though they should be. An argument could be made that with every misguided tweet, Trump is bringing his vice president a step closer to the presidency. Dan Quayle, who is a friend of Pence's from Indiana, has advised Pence to make sure that his staff is integrated with the president's West Wing staff. "He [Pence] is very focused, you don't see a lot of emotion. He's a very matter-of-fact, get-the-job-done kind of guy," Quayle said. "I think that's probably one of the reasons Trump picked him." At his first Cabinet meeting Trump looked around the room and asked, "Where is our vice president?" Pence, ever the happy warrior, was seated right in front of him. Pence then testified to Trump's superiority, telling the assembled Cabinet secretaries (and the reporters in the room): "The greatest privilege of my life is to serve as the vice president to the president who's keeping his word to the American people and assembling a team that's bringing real change, real prosperity, real strength back to our nation."

Trump's legislative director, Marc Short, had worked for the powerful political network overseen by billionaire Charles G. Koch. Charles and his brother David are known for their support of conservative causes and their billionaire-led network of donors and they are Pence allies. Short said that during efforts to repeal and replace the Affordable Care Act, Pence was expected to persuade his former congressional colleagues on how to vote. "We're leaning on him heavily here right now," Short said. But there is a limit to how much Pence can do on the Hill since the man he serves has

alienated so many members of Congress. After Trump attacked individual Freedom Caucus members and in several tweets called for
their defeat in 2018 after they refused to vote for a new healthcare
bill, Pence had to face his former congressional colleagues. They
asked him to stop the president from targeting them. The conservative members hoped Pence would be sympathetic since he was
once a conservative congressman who did not always support President George W. Bush. "I'll pass that along to the president," he
told them, offering no apology or explanation. Notably, the tweets
targeting individual members stopped for a while.

Pence needs his old friends in Congress as much as they need
him. He leaned heavily on Republican allies, including House
Speaker Paul Ryan and Texas congressman Jeb Hensarling. "He sat
where we sit and he understands how the slog works over here. He
knows the realities of legislating," said Brendan Buck, who works
for Speaker Ryan. "What matters is whether there's someone who
can understand what we're saying and clearly communicate that
to the president." Pence and Ryan served together for twelve years
in the House, and they went to Bible study together. When they
meet now it's often last minute, when Pence is on the Hill and
he drops by Ryan's office. Republican congressional leaders view
Pence as an ally. Friends on the Hill describe Pence as "thoughtful"
and "serious," and one senior aide to a top Republican member of
Congress said Pence is their greatest hope to stabilize the Trump
White House and get legislation passed.

Pence has let it be known that he's interested in running for president in 2020 if Trump decides not to run again, and the key will be
fund-raising. Pence is comfortable with the megarich and he is not
intimidated by money, even though he has very little of it himself.
He once gave a friend running for office advice: "They've all got
big egos and they're rich. You've just got to go tell them they don't
need those thousands of dollars, but you do." He has hosted several
dinners with big donors at the Naval Observatory, and during the
2016 campaign he won Trump the support of billionaire industrialist

Charles Koch, who did not support Trump during the campaign and sidelined his organization's enormous political machine in protest. (Charles Koch once said the choice between Trump and Hillary Clinton was like choosing between cancer and a heart attack.) The Koch brothers co-own almost all of Koch Industries, the second-largest private company in the United States. Former White House chief strategist Steve Bannon had cutting words to say about the influence Charles and David Koch would have over Pence if he ever became president. "I'm concerned he'd be a president that the Kochs would own," he said. As vice president, Pence has been courting scores of the country's most influential donors, corporate executives, and conservative political leaders in a series of private gatherings and one-on-one conversations, among them brokerage firm founder Charles Schwab and hedge fund titan Kenneth C. Griffin. During dinners at the Observatory, Pence usually hosts a cocktail hour (though he does not drink). Nursing an Arnold Palmer or a glass of water, he tells the assembled group about the history of the residence, recounts the administration's successes, and takes questions. Guests sit at a few small tables and Pence comes to each table to chat with each guest for a few minutes. At larger events he keeps a chair empty at every table so that he can make his way around the room.

During the campaign, Trump called Republican politicians competing with him for the nomination "puppets" for attending the Kochs' so-called seminars, what the brothers call their secretive fund-raising meetings for the wealthiest conservative donors. One member of the Koch brothers' inner circle said that Pence is "aligned with our network." Texas businessman and influential Republican donor Doug Deason is a Koch ally who has been to those dinners at the Observatory. Even though Pence has fallen out of touch with some of his best friends in Indiana, he has made sure to stay close to donors like Deason, who wield enormous power because of their wealth (Deason's father, Darwin, sold the data-processing company he founded to Xerox in 2010 for $6.4 billion.)

Deason recalls the incongruous sight on election night of bil-

lionaire donors, himself included, at a reception for VIPs upstairs at
the Hilton Hotel in midtown Manhattan, while downstairs Trump
delivered an acceptance speech to his populist supporters. "The
forgotten men and women of our country will be forgotten no
longer," Trump said. The donors celebrating upstairs, including
corporate investor Carl Icahn and David Koch, who is the wealthiest resident of New York City, were hopeful that Trump's victory
would bring less government regulation and lower tax rates. Many
of them would not have been there if it were not for Mike Pence.

Deason says he has no doubt that Pence is positioning himself
to run for president, and even though he praises Trump he clearly
favors Pence. "Mike is that voice that whispers in the president's
ear that keeps him focused and calm," Deason said. "If Mike wasn't
there, it would be a lot more caustic." Deason said that even Donald Trump Jr. wishes his father would hold back on some of his
tweets. "He says something presidential and then he comes back
with a tweet that goes too far."

Trump White House aides like to say that Pence is the yin to
Trump's yang—one Pence aide likened Pence to Felix and Trump
to Oscar (referring to the 1970s television show *The Odd Couple*
about a mismatched pair of lovable roommates, one a neat freak
and the other a slob). "You need to keep your arms and legs in
the ride at all times," Pence told a group of students at American University last year. "Put the roll bar down, because you just
got to hang on." And hanging on is precisely what Pence intends
to do at all costs. One Pence staffer admitted the obvious: "Of
course he's looking ahead, but he knows he needs this administration to succeed." Where Trump threatens, Pence cajoles, where
Trump issues dire warnings about "American carnage," Pence de-
livers a more inspirational message. He recognizes that he does not
have the same charisma as Trump, nor can he fill stadiums like
Trump can—but he also does not crave adoration from crowds
like Trump does. Unlike Cheney, who had no interest in the presidency when he was vice president, when Pence goes to the Hill to

"touch gloves," as he says, on a weekly basis, he insists on walking through the Capitol Rotunda so that tourists can get their photos taken with him.

Greg Pence, who is about three years older than his brother and looks so much like him that he sometimes gets mistaken for him, said his brother likes the privacy of the Naval Observatory. But he sounded worried about his little brother. "He looks tired," he said, "he looks very tired all the time." Other family members are worried he is losing weight. The strain of working ninety to a hundred hours a week is showing. "I went out there [to Washington] and we watched the NCAA tournament and he fell asleep in the chair while we're watching it because he's so tired."

Pence must be exhausted. He is in the unusual and unenviable position of defending and toning down his boss's words rather than going on the offense, as vice presidents have traditionally done. The vice president's calm demeanor is reassuring to Republicans who worry that Trump is doing major damage to the party. They fear that Trump makes a lot of decisions very quickly and with very little information. Aides have been instructed to keep Trump's briefing papers to one or two pages because he does not have the attention span, or the desire, to read through the vast briefing books given to his predecessors. "He listens," said one very well-known former Republican politician who knows Trump. "I don't know if he absorbs, but he does listen."

They also worry that Jared Kushner, Trump's son-in-law, who is a White House confidant and intensely loyal to Trump, has a loose grasp of how government works and is not interested in learning. One Republican who has worked in the White House and knows Kushner described his understanding of Washington and political realities as "childish," and said that in private meetings Kushner is brimming with a false sense of confidence. (In the early days of the administration, when asked who his father-in-law was bringing in to lead top government agencies—some of which Kushner himself couldn't name or explain—he said, "Don't worry, we're bringing in billionaires.") When Dan Coats, the director of national intelligence who has nearly two decades of service in the House and the Senate, brought up a highly sensitive national security issue with the president, Trump told him to "go talk to Jared about that." It was a little jarring to Coats to be told to talk to a thirty-six-year-old with no government experience, according to a person familiar with the discussion. Kushner and Trump's eldest daughter, Ivanka, who also has an official title and office in the White House, occasionally stop by Pence's West Wing office to ask for advice.

Bill Smith, Pence's former chief of staff, said: "There's no doubt that he's brought in on the inner circle, but the question is there are several inner circles. He was obviously informed of certain things, but when he found out that he was not being leveled with, you found out a little bit of Mike Pence's strength." Smith was referring to Pence's assertions that he did not know about meetings between Russian officials and Trump's first national security adviser, Michael Flynn. Pence said on television that Flynn had not discussed with then–Russian ambassador Sergey Kislyak the sanctions President Obama had imposed on Russia after its meddling in the 2016 election. Even though people in the Trump administration knew Flynn was under FBI scrutiny, he only resigned when Pence discovered Flynn had lied to him.

Former Indiana attorney general Greg Zoeller wonders how his old friend can square his deeply held Christian beliefs with a president who weaponizes his Twitter account on an almost daily basis. Zoeller said Pence "went into the foxhole with someone he might not be aligned with philosophically or morally" out of a sense of duty. Often Pence first sees the president's tweets when they appear in the daily news summary his staff puts together. And when he reads them, aides insist, he quickly moves on to the task at hand, keeps his head down, and seeks to be the calm in the storm.

Of the thirteen men in this book who've served as vice president, four became president, three received the highest civilian award the country has to offer, and one has been awarded the Nobel Peace Prize. Being vice president does not have to spell the end of a career, or of political influence. While he went on to become president, George H. W. Bush is still engaged in politics and has talked privately with a member of Trump's Cabinet about the nuclear threat posed by North Korea. Most became wealthy after leaving office, including Dan Quayle, who is chairman of the global investment group at Cerberus, a private equity and hedge fund firm that manages more than $20 billion. When Al Gore lost the 2000 election his net worth was well below $2 million, but now a series of lucrative business deals and affiliations has netted him tens of millions. In 2007, he was awarded the Nobel Peace Prize for his work on climate change.

Modern presidents have picked their vice presidents to help them govern, not just to get elected. Bill Clinton, George W. Bush, and Barack Obama have asked their vice presidents to be far more than "standby equipment." Like most presidents, these men each had enormous egos, but Bill Clinton listened when Al Gore pushed for humanitarian action in Bosnia; George W. Bush wanted Dick Cheney to play a decisive role in America's response to 9/11; and

Barack Obama leaned on Joe Biden for gut checks on key issues. Clinton, Bush, and Obama were all born after World War II and Obama was born in 1961, three years off from the "Generation X" label generally applied to those born between 1964 and 1980. They allowed their vice presidents to have more power in part because the job of the president has become infinitely more difficult in the post–World War II era with growing national security threats and looming economic concerns. Unlike Eisenhower, Kennedy, and Johnson, whose relationships with their vice presidents were marked by contempt, or Nixon and Bush, whose relationships with their vice presidents were ones of indifference or annoyance, Clinton, Bush, and Obama were less concerned about outward status and more interested in using their vice presidents effectively. Jimmy Carter and Walter Mondale deserve credit for this shift from a marriage of convenience to a true partnership.

Donald Trump, however, has turned back the clock, and his relationship with Mike Pence is more like Lyndon Johnson's dynamic with Hubert Humphrey. Johnson, like Trump, was also a larger-than-life and deeply polarizing personality. Trump would prefer his VP be seen and not heard and Pence's frequent rejections of interview requests echo Johnson's edict that Humphrey travel with no press: both men knew well the risks of getting more publicity than the presidents they served. While Pence is dispatched to Capitol Hill weekly to try to get legislative issues passed, he has not been handed clear authority over any particular issue, his chief of staff insists that's the way they want it, but it leaves Pence with no job description and no single area that he can take credit for, should he run for president himself one day.

———

A vice president can exercise enormous power while in office, as Dick Cheney did in the first term of George W. Bush's presidency, and as Al Gore and Joe Biden did in Bill Clinton's and Barack Obama's administrations, but they have to establish parameters privately with the presidents they serve. "In meetings President Obama would tend not to put his thumb on the scale because he knew if he said what he thought, everybody else would clam up," says Valerie Jarrett, Obama's senior adviser and close friend. "He's a good listener so he will let people state their positions, and Vice President Biden was never shy about stating his. He stated it without knowing where President Obama stood." That kind of honesty can be difficult, but it is important—being counselor in chief is the best way to exercise real power as vice president.

When I covered the White House for Bloomberg News during the Obama administration, I traveled with Biden from time to time. These trips were nothing like trips on Air Force One, where every press seat was full and access to the president was limited. Traveling with the vice president was a much more relaxed affair, and sometimes only one or two reporters tagged along. There was usually a chance to chat with the VP, who almost always seemed up for it. "Anyway, I talk too much," Biden said once, after a long chat, giving a sideways glance to his press aide. "These guys are going to get very nervous."

Biden gets emotional when he talks about his friendship with Obama. Obama tells people that Biden is like his older brother, he can "say things that other people can't say." And Biden says that they made up for each other's weaknesses, but is quick to add: Obama "made up for a hell of a lot more of mine than I did his." One way Biden helped Obama had nothing to do with politics—he gave him a family he never had.

Since leaving the vice presidency, Biden spends so much time traveling that he is rarely in his Washington, D.C., office. His desk is bare except for a few framed photos of his family. A photo hangs on the wall from election night ten years ago—it was taken in Chicago's Grant Park, and in the photo Obama is beaming as Biden's mother, Jean, then ninety-one years old, reaches for Obama's hand as they walk triumphantly onstage with Biden before an adoring crowd of an estimated 240,000 people. During our meeting Biden led me over to the photo hanging on the wall. He pointed to his mother, who passed away in 2010. "She wasn't supposed to walk out," he said. "And she says, 'Come on, honey, it's going to be OK,' to Barack." Think of it, Biden said, his mother was trying to reassure the newly elected president. Looking at the photo knowing this, Obama looks less like a global superstar and more like a vulnerable man grateful for the reassurance.

"I joke with Michelle [Obama] that her mom, Mrs. Robinson, could have been my mother. Michelle and I were raised similarly, my dad was economically a click above, but everything with my dad was always about everybody else first, like her dad. That's the family Barack wanted. He'd ask me questions at lunch that made it clear that he didn't have the benefit of some of what Michelle and I had." Once, Biden said, when he was vice president, Michelle asked him to talk with Obama about something personal. He would not say what it was because it was a private matter.

"I can't," Biden told her.

"Look," Michelle told him, "you're the only one he completely trusts."

"Why?" Biden asked her.

"His experience, Joe, is that everyone who's been around him his whole adult life has wanted something from him. And you've not wanted anything. You've demonstrated that you would jump in front of a train for him, people haven't done that before."

Biden says he likes to think he helped Obama get in touch with his emotions. "Barack is not nearly as openly passionate as I am. Either I'm much more extroverted or he's much more introverted," he said, taking a long pause. "I remember way after the fact, reading how much his eulogy of my son blew away the press. When he said, 'I'm a Biden.' He meant it," Biden said, as he slowly wiped away tears.

———

Dan Quayle was thrown into the vice presidency with little support and he faced a suspicious West Wing staff. But Quayle told his old friend former Indiana attorney general Greg Zoeller there was no reason to feel bad for him, he was vice president of the United States after all. "I think Pence finds himself in that spot," Zoeller said of the other Indiana vice president he knows well. "Even if he just gets beaten to a pulp and never is heard from again and runs off to Siberia, he was vice president of the United States. What's there to be sorry about?"

But when vice presidents try to win the presidency and lose, the defeat can sting even more. Was there any part of Tipper Gore that was relieved after her husband lost the 2000 election? "To be honest, no," she said. There is a sense that something is owed to the vice president who spends years in the background, setting aside his own ambition to serve the president. When Biden said, "Do I regret not being president? Yes," he was speaking for virtually every vice president in history.

Mike Pence models his vice presidency after George H. W. Bush's, not Dick Cheney's—he has tried to position himself as far away from that image as possible. Like Pence, Bush was always sensitive about not wanting to look like he was upstaging his boss (as if anyone could upstage Reagan or Trump). Pence's aides like to point out that Bush, like Pence, served a president who was larger than life and also a Washington outsider. But perhaps there is another, more important and much more honest reason behind Pence's admiration for Bush: Bush did something that no vice president has managed to do since—he became president.

ACKNOWLEDGMENTS

I'd like to thank the former vice presidents, from both parties, who spoke with me so candidly about the obvious opportunities and lesser-known daily humiliations that accompany the vice presidency: Joe Biden, Dick Cheney, Al Gore, Dan Quayle, George H. W. Bush, and Walter Mondale. And Jimmy Carter, whose appreciation for his own vice president, Walter Mondale, is sincere and refreshing in an age when sarcasm and cynicism are pervasive in politics.

The aides, family members, and friends of the vice presidents and the presidents they served wanted to lift the curtain on this often overlooked relationship and give credit to the men who are more often the butt of jokes than the receivers of accolades, among them were: Tipper Gore, Nick Ayers, Dan Coats, Lynda Pedersen, Sarah Eaton, Liz Haenle, Neil Patel, Kate Bedingfield, William Russo, Steve Ricchetti, Ted Kaufman, Mike Donilon, Ron Klain, Margaret Aitken, Sarah Bianchi, Kalee Kreider, Deb Greenspan, Michael Feldman, Carter Eskew, Rob Hamilton, Donald Rumsfeld, Bill Kristol, Bruce Reed, Ken Baer, Valerie Jarrett, David Axelrod, Jay Carney, Josh Earnest, Alyssa Mastromonaco, Jim McGrath, Tom Collamore, Craig Fuller, James Smith, Bess and Tyler Abell, Jim Ketchum, Sid Davis, Al Spivak, Gregory Pence, Jim Atterholt, Jerry Rafshoon, Ed Simcox, Pete Seat, David McIntosh,

Brendan Buck, Kent Sterling, Steve Simpson, Greg Garrison, Jeff Cardwell, Bill Smith, Marc Short, Josh Pitcock, and Marc Lotter. The friends and advisers to Vice President Mike Pence were especially helpful because it is always more difficult to examine a relationship when it is in its earliest stages.

My literary agent, Howard Yoon, is a true partner and he is always in my corner, no matter what, as is Gail Ross, who came up with the idea for this book. My editor at HarperCollins, Gail Winston, cares deeply about the authors she works with and is nurturing, supportive, and talented. Roger Labrie, Sofia Groopman, Emily Taylor, and Victor Hendrickson kept making this book better and better and Sofia is a master of organization.

I appreciate the help of the women who work at the Presidential Materials Division at the National Archives: Stephannie Oriabure, Jessica Owens, and Anna Yallouris. The staffs at several presidential libraries helped me cull through thousands of pages of letters and memos and they include Mary McMurray, Sam Rushay, and David Clark at Harry Truman's Library; Liza Talbot at Lyndon Johnson's library; Scott Russell and Jonathan Movroydis at Richard Nixon's library; and Elaine Didier, John O'Connell, Kate Murray, and Stacy Davis at Gerald Ford's library. And thank you to Jamin Goecker, a student at the Bush School of Government and Public Service at Texas A&M University, for his excellent research at George H. W. Bush's library; Danielle Fowler, a researcher at Ronald Reagan's library; and Rhonda Young at Bill Clinton's library.

This book, like most books, was a family affair. My husband and best friend, Brooke Brower, is a gifted editor who helped me clarify and organize my thoughts. He put up with me during the sometimes frustrating process and he served as a lighthouse, helping to navigate the rocks that inevitably cropped up along the way. And, perhaps best of all, he never forgets to bring me coffee in the morning. My parents, Christopher and Valerie Andersen, also helped every step of the way. My mother is a voracious reader and a brilliant editor and my father is one of the smartest people I've ever met. Our children, Graham and Charlotte, may not have helped with editing (they're four and five years old!) but they did make every single day better and they provided some much-needed comic relief when interviews fell through at the last minute, or when the writing process felt overwhelming. They have helped me see the humanity in everyone, and, like most parents I know, having them has given me a heightened sense of empathy for the personal and professional struggles faced by everyone—even politicians.

I hope I have managed to provide a glimpse into the relationship between the two most powerful people in the country. The bonds between them are, as one aide to Biden put it, "forged in fire," and critical to the future of our country.

Note on Reporting

I spent more than two years working on this book, and I interviewed more than two hundred people, including every living former vice president. My research for *First in Line* took me to Minneapolis, Minnesota, where I interviewed Walter Mondale in his quiet law firm office; to a bustling private equity firm in Manhattan, where I sat down with Dan Quayle; to Dick Cheney's understated home in leafy McLean, Virginia, and to Joe Biden's corner office in downtown Washington, D.C. I also interviewed Al Gore and George H. W. Bush. The candor of all six men was remarkable. The former vice presidents described conflicts with the presidents they served and the touching and deeply personal friendships they sometimes forged with those presidents.

I talked to dozens of senior White House aides who saw firsthand the interactions between the most powerful man in the world and his number two. Many of them recounted events that have not been made public until now. Jimmy Carter, Valerie Jarrett, Dan Coats, Nick Ayers, Bill Kristol, Donald Rumsfeld, David Axelrod, and many others offered insight into the relationship from the president's point of view. I sat down with people who candidly described the vetting process, and others who talked about the inherent tension that comes with the often-infantilizing role of the vice presidency. Most of these conversations were in person and in some circumstances, especially when it came to highly personal anecdotes, sources asked not to be named because of the sensitivity of the subject matter. I have respected their wishes. Firsthand accounts were supplemented with archival materials, including oral histories from presidential libraries, as well as memoirs and biographies.

Prologue: "Safe Hands"
Interview subjects include A. B. Culvahouse, Dan Coats, Nick Ayers, Jim Atterholt, Ed Simcox, David McIntosh, and Gary Varvel.

I. You're a Guest in My House
Interview subjects include Joe Biden, Dick Cheney, Jimmy Carter, Al Gore, George H. W. Bush, David Axelrod, Walter Mondale, Ron Klain, Dan Quayle, Valerie Jarrett, Nick Ayers, Dan Coats, Stuart Spencer, Josh Earnest, Bruce Reed, Donald Rumsfeld, Mike Donilon, Jim Atterholt, Greg Zoeller, Adam Frankel, Carter Eskew, Brendan Buck, Jerry Rafshoon, Adam Frankel, Steve Ricchetti, Richard Moe, and Jamal Simmons. Published material includes: Jules Witcover, "The Dynasty that Almost Wasn't: Jeb's Chances at a family three-peat are slimming, but Bushes were never destined for the White House," *Politico Magazine*, September 15, 2015; Jules Witcover, *Very Strange Bedfellows: The Short and Unhappy Marriage of Richard Nixon and Spiro Agnew* (New York: Public Affairs, 2007); Robert A. Caro, *The Years of Lyndon Johnson: The Passage of Power* (New York: Alfred A. Knopf, 2012); Laurence Burd, "Kennedy Picks Johnson to Be Stand-In," *Chicago Tribune*, August 11, 1961; Interview conducted by Yanek Mieczkowski with Stu Spencer, April 3, 2007, Palm Desert, California, Gerald R. Ford Presidential Library; William Doyle, *Inside the Oval Office: White House Tapes from FDR to Clinton* (New York: Kosansha USA Inc., 1999); Transcript: Vice President Cheney on *FOX News Sunday*, December 22, 2008; Dan Zak, "Nervous about nukes again? Here's what you need to know about The Button (There is no button)," *Washington Post*, August 3, 2016; David Rutz, "Biden Again Points Out Military Aide Carrying Nuclear Codes at Clinton Rally," *Free Beacon*, September 1, 2016; Interview conducted by Yanek Mieczkowski with President Gerald R. Ford, January 24, 2002, Telephone Interview, President Gerald R. Ford Presidential Library; Joel Goldstein, *The White House Vice Presidency: The Path to Significance, Mondale to Biden* (Lawrence: University Press of Kansas, 2016); Alvin S. Felzenberg, "The Vice Presidency Grows Up," Hoover Institution, February 1, 2001; Interview with Vice President Richard Nixon, Hartmann Papers, Box 33, Gerald R. Ford Presidential Library; Paul Jones interview with President Gerald R. Ford, 1984, Gerald R. Ford Presidential Library; Richard V. Allen, "George Herbert Walker Bush: The Accidental Vice President," *New York Times Magazine*, July 30, 2000; Lady Bird Johnson, *A White House Diary* (New York: Holt, Rinehart and Winston, 1970); Charles O. Jones, *The American Presidency: A Very Short Introduction* (New York: Oxford University Press, 2007);

Alexander Burns, "Prior VP nominee selection dates," *Politico,* August 18, 2008; David S. Broder, "The Triple-H Brand on the Vice-Presidency," *New York Times,* December 6, 1964; "Mrs. Kennedy 'Reclaims' Chandelier from Senate," *Chicago Tribune,* June 23, 1962; Ed Henry, "Cheney to Emanuel: Control your VP," December 17, 2008, http://politicalticker.blogs.cnn .com/2008/12/17/cheney-to-emanuel-control-your-vp/; Peter Baker, *Days of Fire: Bush and Cheney in the White House* (New York: Random House, 2013).

II. Two Men, Two Hotel Suites

Interviews include Dick Cheney, Donald Rumsfeld, Bess and Tyler Abell, Bob Dole, Richard Moe, Ron Klain, Craig Fuller, and Stuart Spencer. Published material includes: Robert A. Caro, *The Years of Lyndon Johnson: The Passage of Power* (New York: Alfred A. Knopf, 2012); Ben Cosgrove, "Head to Head: JFK and RFK, Los Angeles, July 1960," *Time,* May 24, 2014; Horace Busby Oral History Interview VIII, 4/2/89, by Michael L. Gillette, LBJ Presidential Library; Drew Pearson, *Washington Merry-Go-Round: The Drew Pearson Diaries, 1960–1969* (Lincoln: Potomac Books, an imprint of University of Nebraska Press, 2015); James M. Cannon Research Interviews and Notes, Box One, Ford-Cannon Interview, Rancho Mirage, CA 4/29/1990, Gerald R. Ford Presidential Library; Michael Beschloss, "Kennedy, L.B.J. and a Disputed Deer Hunt," *New York Times,* August 15, 2014; Initial Ford-Cannon Interview, Beaver Creek, CO 9/1/1989, Gerald R. Ford Presidential Library; Initial Ford-Cannon Interview, Rancho Mirage, CA 4/26/1990, Gerald R. Ford Presidential Library; Initial Ford-Cannon Interview, Rancho Mirage, CA 4/27/1990, Gerald R. Ford Presidential Library; Thomas Oliphant with Curtis Wilkie, *The Road to Camelot: Inside JFK's Five-Year Campaign* (New York: Simon & Schuster, 2017); James M. Cannon Research Interviews and Notes, 1989–94, Box Two, Folder title: Notes on the Ambrister Interviews (pages 601–700), Gerald R. Ford Presidential Library; James M. Cannon Research Interviews and Notes, An Interview with Robert Hartmann Conducted by James M. Cannon, June 19, 1991, Gerald R. Ford Presidential Library; James M. Cannon Research Interviews and Notes, An Interview with Michael J. "Mike" Mansfield Conducted by James M. Cannon, May 31 1991, Gerald R. Ford Presidential Library; James M. Cannon Research Interviews and Notes, An Interview with William Seidman, Conducted by James M. Cannon, October 2, 1989, Gerald R. Ford Presidential Library; Nancy Gibbs and Michael Duffy, *The Presidents Club* (New York: Simon & Schuster, 2012); James M. Cannon

Research Interviews and Notes, Interview with Peter Rodino Conducted by James M. Cannon, September 28, 1990, Gerald R. Ford Presidential Library; James Cannon, *Time and Chance: Gerald Ford's Appointment with History* (Ann Arbor: University of Michigan Press, 1998); Jon Meacham, *Destiny and Power: The American Odyssey of George Herbert Walker Bush* (New York: Random House, 2015).

III. The Art of the Vet

Interviews include Dick Cheney, Joe Biden, Al Gore, Dan Coats, Donald Rumsfeld, Greg Pence, Dan Quayle, Kent Sterling, Lucy Tutwiler, Jeff Cardwell, Jim Atterholt, Joe Trippi, Bess and Tyler Abell, Jim Johnson, James Hamilton, A. B. Culvahouse, Bob Dole, Bill Kristol, Richard Moe, Ron Klain, Bill Smith, John Weaver, Craig Fuller, Stuart Spencer, Ed Simcox, and Dylan Loewe. Published material includes: Jacqueline Alemany, "Donald Trump Offered Chris Christie Vice President Role Before Mike Pence, Sources Say," CBS News, October 30, 2016, https://www.cbsnews.com /news/donald-trump-offered-chris-christie-vice-president-role-before -mike-pence/; Ashley Parker, Alexander Burns, and Maggie Haberman, "A Grounded Plane and anti-Clinton Passion: How Mike Pence Swayed the Trumps," *New York Times*, July 16, 2016; "Mike Pence on Donald Trump: 'I Believe in Forgiveness,'" CBS News, October 10, 2016, https://www .cbsnews.com/news/mike-pence-on-donald-trump-pence-i-believe-in -forgiveness/; Edward-Isaac Dovere, "The Man Plotting Biden 2016: Steven Ricchetti is the man behind the vice president and possible late-entry presidential candidate," *Politico*, September 2, 2015; Matea Gold and Rosalind S. Hedlerman, "Gingrich is a vice-presidential finalist and his last campaign is still millions in debt," *Washington Post*, July 13, 2016; Marisa M. Kashino, "A.B. Culvahouse: The Man Who Vetted Sarah Palin," *Washingtonian*, April 21, 2011; Evelyn Lincoln, *Kennedy & Johnson* (New York: Holt, Rinehart and Winston, 1968); Judy Klemesrud, "Amy, a Glenn Supporter, Cheers Up in Zoo Visit," *New York Times*, July 16, 1976; Jason Zengerle, "Wanna Be Veep? Okay, but This Is Going to Hurt," *GQ*, July 18, 2012; Peter Baker, *Days of Fire: Bush and Cheney in the White House* (New York: Random House, 2013); Jane Mayer, "The Danger of President Pence," *The New Yorker,* October 23, 2017; Ryan Lizza, "Biden's Brief," *The New Yorker*, October 20, 2008; Matthew Rees, "Al Gore, Midnight Toker," *The Weekly Standard*, February 7, 2000; James M. Cannon Research Interviews and Notes, Box One, Nixon responses to Cannon's questions, Gerald R. Ford Presidential Library; David M. Oshinsky, "Fear and Loathing in the White House,"

New York Times, October 26, 1997; Dan Quayle, *Standing Firm* (New York: HarperCollins, 1994); Bob Woodward and David S. Broder, *The Man Who Would Be President* (New York: Simon and Schuster, 1992); Philip Bump, "Sarah Palin cost John McCain 2 million votes in 2008, according to a study," *Washington Post*, January 19, 2016; Valerie J. Nelson, "Ethel Bradley dies at 89; wife of Los Angeles Mayor Tom Bradley," *Los Angeles Times*, November 26, 2008; Jonathan Karl, "Dick Cheney: Picking Sarah Palin for VP Was 'A Mistake,'" ABC News, July 29, 2012; Evan Osnos, "The Biden Agenda: Reckoning with Ukraine and Iraq, and keeping an eye on 2016," *The New Yorker*, July 28, 2014; Ben Terris, "Sherrod Brown thinks he could have helped Democrats win in 2016. But what about 2020?," *Washington Post*, July 24, 2017; Joe Biden, *Promises to Keep* (New York: Random House, 2007); Spiro Agnew, *Go Quietly . . . or else* (New York: William Morrow and Company, Inc., 1980); "Richard Nixon, John D. Ehrlichman, and H. R. 'Bob' Haldeman on 20 July 1971," Conversation 263–009 (*PRDE* Excerpt A), *Presidential Recordings Digital Edition* ["Vice President Agnew," ed. Nicole Hemmer] (Charlottesville: University of Virginia Press, 2014), http://prde.upress.virginia.edu/conversations/4004759; Nicole Hemmer, "Why the Vice Presidency Matters," *The Atlantic*, July 21, 2016; Jon Meacham, *Destiny and Power: The American Odyssey of George Herbert Walker Bush* (New York: Random House, 2015); Michael Wolff, "Full Bloom: Sexual-harassment lawsuits have torpedoed plenty of political careers. So why has the media given Mike Bloomberg a pass on the charges made against him," *New York Magazine,* November 5, 2001; Alan Feuer, "54 More Women Accuse Bloomberg Firm of Bias," *New York Times*, May 2, 2008; Maureen Dowd, "Biden's Debate Finale: An Echo From Abroad," *New York Times*, September 12, 1987; Philip Elliot, "The Persistent Passion of Vice President Mike Pence," *Time,* May 25, 2017; Turner Cowles and Michael Barbaro, "What Mike Pence Brings to Donald Trump's Campaign," *New York Times*, July 15, 2016; Tom Alberta, "Man on a Wire: Mike Pence's Tightrope Act," *Politico*, July/August 2017; Rafi Letzter, "Mike Pence, the vice president of the United States, has said he doesn't believe that smoking kills," *Business Insider*, January 21, 2017; Nolan D. McCaskill, "Pence shows Trump how to sweat it out with Texas victims," *Politico*, August 31, 2017.

IV. The Observatory

Interview subjects include Joe Biden, Dan Quayle, Tipper Gore, Barbara Bush, Walter Mondale, Dick Cheney, Paula Trivette, Craig Fuller, Katie McCormick Lelyveld, Elizabeth Haenle, and Carlos Elizondo. Published material includes: Lynne Cheney, "This Is Where Many Vice Presidents Have Lived," *Architectural Digest*, December, 2001; Jura Koncius, "As the Bidens pack up, a look at their mark on the vice president's residence," *Washington Post*, January 11, 2017; Jane Mayer, "The Danger of President Pence," *The New Yorker,* October 23, 2017; Maureen Groppe, "The secret history of Mike Pence's new house," *USA Today*, November 24, 2017; Peter Baker, *Days of Fire: Bush and Cheney in the White House* (New York: Random House, 2013).

V. The Second Lady

Interview subjects include Tipper Gore, Barbara Bush, Dick Cheney, Laura Bush, Dan Quayle, Valerie Jarrett, Marguerite Sullivan, Pete Seat, Elizabeth Haenle, Jamal Simmons, Senator Roy Blunt, Paula Trivette, Carlos Elizondo, Ed Simcox, Trooper Sanders, Neil Patel, Jim Atterholt, and Elaine Kamarck. Published material includes: Thomas M. DeFrank, *Write It When I'm Gone: Remarkable Off-the-Record Conversations with Gerald R. Ford* (New York: G.P. Putnam's Sons, 2007); Jeff Horwitz and Julie Bykowicz, "Still No Charity Money from Leftover Trump Inaugural Funds," Associated Press, September 15, 2017; Ashley Parker, "Karen Pence is the vice president's 'prayer warrior,' gut check and shield," *Washington Post*, March 28, 2017; Barbara Bush, *Barbara Bush: A Memoir* (New York: Scribner, 1994); John M. Broder, "Clinton's Affair Took a Toll on Relationship with Gore," *Washington Post*, March 3, 2000; Laura Bush, *Spoken from the Heart* (New York: Simon & Schuster, 2010); Marjorie Williams, "The Marilyn Quayle Factor," *Washington Post*, October 20, 1988; Danielle Gay, "Let's talk about Melania Trump's 25 carat White House portrait diamond," *Vogue*, April 4, 2017; Emma Kate Fittes, "Karen Pence has BSY grad make inauguration gowns," *Indianapolis Star*, January 19, 2017; Miller Center, "Interview with Elaine Kamarck," University of Virginia, May 7–8, 2008, February 13, 2017, http://millercenter.org/oralhistory/interview/elaine-kamarck; Ceci Connolly, "Tipper Gore Details Depression Treatment," *Washington Post*, May 8, 1999; Lloyd Grove, "Tipper Gore, Al Gore Separation: Why They Split," *The Daily Beast*, June 2, 2010; "Lynne Cheney novel churns controversy in Senate race," CNN, October 30, 2006; Karen Gullo, "Vice President's Wife Back at Work," Associated Press, January 22, 2001; Shari Rudavsky, "Karen Pence is right at home," *Indianapolis Star*, December 12, 2013; Melissa Langsam Braunstein, "Second Lady Karen Pence Opens Up About Her Struggles with Infertility," thefederalist.com; Alexandra Macon, "Joe and Jill Biden's Love Story Will Pull at Your Heartstrings," *Vogue*, November 17, 2016;

Jane Mayer, "The Danger of President Pence," *The New Yorker,* October 23, 2017; Virginia Coyne, "The Vice President's Residence," *Washington Life,* December 2017; Melinda Henneberger, "This Second Lady Is Keeping Her Day Job," *New York Times,* February 6, 2001; James Rosen, *Cheney One on One: A Candid Conversation with America's Most Controversial Statesman* (New York: Regnery, 2015); David Maraniss and Ellen Nakashima, *The Prince of Tennessee: The Rise of Al Gore* (New York: Simon & Schuster, 2000).

VI. Tragedy and Trauma

Interviews include George H. W. Bush, Jimmy Carter, Bess and Tyler Abell, Luci Baines Johnson, Steve Ford, Connie Stuart, Sid Davis, Susan Ford, and Al Spivak. Published material includes: Harry Truman, *1945: Year of Decisions: Memoirs, Volume 1* (New York: New Word City, 2014); A. J. Baime, *The Accidental President: Harry S. Truman and the Four Months that Changed the World* (New York: Houghton Mifflin Harcourt, 2017); Robert A. Caro, *The Years of Lyndon Johnson: The Passage of Power* (New York: Alfred A. Knopf, 2012); Jan Jarboe Russell, "Lady Bird Looks Back," *Texas Monthly,* December 1994; Jack Valenti, "The President Is Dead, You Know," *Texas Monthly,* June 2007; Stewart Alsop, *The Center: People and Power in Political Washington* (New York: Popular Library, 1968); Jules Witcover, *Very Strange Bedfellows: The Short and Unhappy Marriage of Richard Nixon and Spiro Agnew* (New York: Public Affairs, 2007); Adam Clymer, "Thomas F. Eagleton, 77, a Running Mate for 18 Days, Dies," *New York Times,* March 5, 2007; Initial Ford-Cannon Interview, Beaver Creek, CO 9/1/1989, Gerald R. Ford Presidential Library; Initial Ford-Cannon Interview, Rancho Mirage, CA 4/26/1990, Gerald R. Ford Presidential Library; Initial Ford-Cannon Interview, Rancho Mirage, CA 4/27/1990, Gerald R. Ford Presidential Library; James M. Cannon Research Interviews and Notes, 1989–94, Box two, Folder title: Notes on the Ambrister Interviews (pages 601–700), Gerald R. Ford Presidential Library; James M. Cannon Research Interviews and Notes, An Interview with Robert Hartmann Conducted by James M. Cannon, June 19, 1991, Gerald R. Ford Presidential Library; James M. Cannon Research Interviews and Notes, An Interview with Michael J. "Mike" Mansfield Conducted by James M. Cannon, May 31, 1991, Gerald R. Ford Presidential Library; James M. Cannon Research Interviews and Notes, An Interview with William Seidman, Conducted by James M. Cannon, October 2, 1989, Gerald R. Ford Presidential Library; James M. Cannon Research Interviews and Notes, An Interview with Peter Rodino Conducted by James M. Cannon, September 28, 1990, Gerald R. Ford Presidential

Library; James Cannon, *Time and Chance: Gerald Ford's Appointment with History* (Ann Arbor: University of Michigan Press, 1998); Spiro Agnew, *Go Quietly . . . or else* (New York: William Morrow and Company, Inc., 1980); "Richard Nixon, John D. Ehrlichman, and H. R. 'Bob' Haldeman on 20 July 1971," Conversation 263–009 (*PRDE* Excerpt A), *Presidential Recordings Digital Edition* ["Vice President Agnew," ed. Nicole Hemmer] (Charlottesville: University of Virginia Press, 2014), http://prde.upress.virginia.edu/conversations/4004759; Nicole Hemmer, "Why the Vice Presidency Matters," *The Atlantic*, July 21, 2016; Television Report, November 27, 1969, 11/26 Telecasts, President's Office Files, Box 31, Richard Nixon Presidential Library; Talking Points for Agnew, July 21, 1972, President's Personal File, Box 5, Richard Nixon Presidential Library; Letter from President Richard Nixon to Spiro Agnew, October 29, 1972, President's Personal File, Box 5, Richard Nixon Presidential Library; Doris Kearns Goodwin, *Lyndon Johnson and the American Dream* (New York: St. Martin's Press, 1976); Chris Jones, "The Flight from Dallas," *Esquire*, September 16, 2013; Chris Jones, "The Most Incredible Documents from JFK's Assassination," *Esquire*, November 22, 2013; Frank Carlucci Files, Ronald Reagan Presidential Library; Briefing for the Press by Secretary Haig, March 30, 1981, Ronald Reagan Presidential Library; Richard V. Allen Papers, Ronald Reagan Presidential Library; Jon Meacham, *Destiny and Power: The American Odyssey of George Herbert Walker Bush* (New York: Random House, 2015).

VII. From Senator to Subordinate: The Story of Nixon/Eisenhower, Johnson/Kennedy, and Humphrey/Johnson

Interview subjects include Jimmy Carter, Richard Moe, Ron Klain, Tyler and Bess Abell, Adam Frankel, Joe Califano, and Walter Mondale. Published material includes: President's Personal File, Box 47, March 1969, Nixon's handwritten notes re: Ike Meetings, Richard Nixon Presidential Library; Vice-Presidential Collection-PPS 325-Box 3, June 16, 1953, Memorandum on the duties of the vice president, Richard Nixon Presidential Library; Thomas Mallon, "Wag the Dog," *The New Yorker*, February 4, 2013; Vice-Presidential Collection-PPS 324-Box 1, October 1, 1955, Memorandum on the duties of the vice president, Richard Nixon Presidential Library; Pre-Presidential Papers, Series 320, Box 237, Eisenhower, Dwight D., Richard Nixon Presidential Library; Drew Pearson, *Washington Merry-Go-Round: The Drew Pearson Diaries, 1960–1969* (Lincoln: Potomac Books, an imprint of University of Nebraska Press, 2015); Transcript: Vice President Cheney on *FOX News Sunday*, December 22, 2008; Robert A. Caro, "The Transition: Lyndon Johnson and the events in Dallas," *The New Yorker*, April 2, 2012; Dan Zak, "Nervous about nukes again? Here's what you need to know about The Button (There is no button)," *Washington Post*, August 3, 2016; Robert A. Caro, *The Years of Lyndon Johnson: The Passage of Power* (New York: Alfred A. Knopf, 2012); Doris Kearns Goodwin, *Lyndon Johnson and the American Dream* (New York: St. Martin's Press, 1976); Barron J. Lerner, "An M.D.'s Guide to Ike's Heart and Hearth," *New York Times*, January 13, 2004; "Lyndon Johnson And Everett Dirksen On 31 October 1968," http://prde.upress.virginia.edu/conversations/4006113/Notes_Open, *Presidential Recordings Digital Edition* [Chasing Shadows, ed. Ken Hughes] (Charlottesville: University of Virginia Press, 2014); "Lyndon Johnson and Hubert Humphrey on 31 October 1968," http://prde.upress.virginia.edu /conversations/4006117, *Presidential Recordings Digital Edition* [Chasing Shadows, ed. Ken Hughes] (Charlottesville: University of Virginia Press, 2014); Thomas Oliphant with Curtis Wilkie, *The Road To Camelot: Inside JFK's Five-Year Campaign* (New York: Simon & Schuster, 2017); Jeffrey Frank, *Ike and Dick: Portrait of a Strange Political Marriage* (New York: Simon and Schuster, 2013); Carl Solberg, *Hubert Humphrey: A Biography* (New York: Borealis Books, 1984).

VIII. Confusion, Conflict, and Musical Chairs: The Rocky Road of Agnew/Nixon, Ford/Nixon, and Rockefeller/Ford

Interview subjects include Jimmy Carter, Dick Cheney, Bob Dole, Donald Rumsfeld, Richard Moe, Susan Ford, Steve Ford, Ron Klain, and Walter Mondale. Published material includes: President's Personal File, Box 47, March 1969, Nixon's handwritten notes re: Ike Meetings, Richard Nixon Presidential Library; Vice-Presidential Collection-PPS 325-Box 3, June 16, 1953, Memorandum on the duties of the vice president, Richard Nixon Presidential Library; Vice-Presidential Collection-PPS 324-Box 1, October 1, 1955, Memorandum on the duties of the vice president, Richard Nixon Presidential Library; Pre-Presidential Papers, Series 320, Box 237, Eisenhower, Dwight D., Richard Nixon Presidential Library; Drew Pearson, *Washington Merry-Go-Round: The Drew Pearson Diaries, 1960–1969* (Lincoln: Potomac Books, an imprint of University of Nebraska Press, 2015); Transcript: Vice President Cheney on *FOX News Sunday*, December 22, 2008; Dan Zak, "Nervous about nukes again? Here's what you need to know about The Button (There is no button)," *Washington Post*, August 3, 2016; Barron J. Lerner, "An M.D.'s Guide to Ike's Heart and Hearth," *New York Times*, January 13, 2004; Jules Witcover, *Very Strange Bedfellows: The Short and Unhappy Marriage of Richard Nixon and Spiro Agnew* (New York: Public Affairs, 2007); Adam Clymer, "Thomas F. Eagleton, 77, a Running Mate for 18 Days, Dies," *New York Times*, March 5, 2007; Initial Ford–Cannon Interview, Beaver Creek, CO 9/1/1989, Gerald R. Ford Presidential Library; Initial Ford–Cannon Interview, Rancho Mirage, CA 4/26/1990, Gerald R. Ford Presidential Library; Initial Ford–Cannon Interview, Rancho Mirage, CA 4/27/1990, Gerald R. Ford Presidential Library; James M. Cannon Research Interviews and Notes, 1989–94, Box two, Folder title: Notes on the Ambrister Interviews (pages 601–700), Gerald R. Ford Presidential Library; James M. Cannon Research Interviews and Notes, An Interview with Robert Hartmann Conducted by James M. Cannon, June 19, 1991, Gerald R. Ford Presidential Library; James M. Cannon Research Interviews and Notes, An Interview with Michael J. "Mike" Mansfield Conducted by James M. Cannon, May 31 1991, Gerald R. Ford Presidential Library; James M. Cannon Research Interviews and Notes, An Interview with William Seidman, Conducted by James M. Cannon, October 2, 1989, Gerald R. Ford Presidential Library; James M. Cannon Research Interviews and Notes, An Interview with Peter Rodino, Conducted by James M. Cannon, September 28, 1990, Gerald R. Ford Presidential Library; James Cannon, *Time and Chance: Gerald Ford's Appointment with History* (Ann Arbor: University of Michigan Press, 1998); Spiro Agnew, *Go Quietly . . . or else* (New York: William Morrow and Company, Inc., 1980); "Richard Nixon, John D. Ehrlichman, and H. R. 'Bob' Haldeman on 20 July 1971," Conversation 263–009 (*PRDE* Excerpt A), *Presidential Recordings Digital Edition* ["Vice President Agnew," ed. Nicole Hemmer] (Charlottesville: University of Virginia Press, 2014), http://prde.upress.virginia.edu/conversations /4004759; Nicole Hemmer, "Why the Vice Presidency Matters," *The Atlantic*, July 21, 2016; Television Report, November 27, 1969, 11/26 Telecasts, President's Office Files, Box 31, Richard Nixon Presidential Library; Talking Points for Agnew, July 21, 1972, President's Personal File, Box 5, Richard Nixon Presidential Library; Letter from President Richard Nixon to Spiro Agnew, October 29, 1972, President's Personal File, Box 5, Richard Nixon Presidential Library.

IX. Getting to Know You . . . or Not: Mondale/Carter, Bush/ Reagan, and Quayle/Bush

Interview subjects include Jimmy Carter, Dick Cheney, Bill Kristol, Walter Mondale, Stuart Spencer, Dan Quayle, Craig Fuller, Greg Zoeller, Joe Trippi, Richard Moe, Christine Limerick, Tom Collamore, and Ron Klain.

Published material includes: Jon Meacham, *Destiny and Power: The American Odyssey of George Herbert Walker Bush* (New York: Random House, 2015); Dan Quayle, *Standing Firm: A Vice-Presidential Memoir* (New York: Harper-Collins, 1994); Ronald Reagan, *The Reagan Diaries* (New York: Harper-Collins, 2007); Joel Goldstein, *The White House Vice Presidency: The Path to Significance, Mondale to Biden* (Lawrence: University Press of Kansas, 2016); An Oral History with Walter Mondale, Interviewer: Anita Hecht, Recording Date: May 13, 2009, Wisconsin Historical Society; David Hoffman, "Bush Splits with Reagan on Handling of Noriega," *Washington Post*, May 19, 1988; Richard V. Allen, "George Herbert Walker Bush: The Accidental Vice President," *New York Times Magazine*, July 30, 2000.

X. From Friendship to Betrayal: The Breakup of Al Gore and Bill Clinton

Interview subjects include Al Gore, Ron Klain, Carter Eskew, Michael Feldman, Tipper Gore, Sarah Bianchi, Bruce Reed, Elaine Kamarck, Ken Baer, and Jamal Simmons. Published material includes: Thomas M. DeFrank, *Write It When I'm Gone: Remarkable Off-the-Record Conversations with Gerald R. Ford* (New York: G.P. Putnam's Sons, 2007); Robert G. Kaiser and John F. Harris, "Shalala's Remarks Irk President," *Washington Post*, September 11, 1998; Taylor Branch, *The Clinton Tapes: Wrestling History with the President* (New York: Simon & Schuster, 2009); David Remnick, "The Wilderness Campaign: Al Gore lives on a street in Nashville," *The New Yorker*, September 13, 2004; Ronald Brownstein, "The Times Poll: Bush, Perot in Near Tie with California Voters," *Los Angeles Times,* April 28, 1992; David Maraniss and Ellen Nakashima, *The Prince of Tennessee: The Rise of Al Gore* (New York: Simon & Schuster, 2000); David Maraniss and Ellen Nakashima, "Al Gore, Growing Up in Two Worlds," *Washington Post*, October 10, 1999; John M. Broder, "Clinton's Affair Took a Toll on Relationship with Gore," *Washington Post*, March 3, 2000; "Huma Abedin on Clinton-Gore relations: 'No love lost,'" *Politico*, October 21, 2016; Miller Center, "Interview with Elaine Kamarck," University of Virginia, May 7–8, 2008 http://millercenter.org/oralhistory/interview/elaine-kamarck; Presidential Daily Diary, 12/13/00, Clinton Presidential Library; Carter Eskew, "Gore knew how to lose gracefully," *Washington Post*, October 23, 2016; John Heilemann, "The Comeback Kid," *New York Magazine*, May 29, 2006; Jodi O'Connell, "Where are they now: What's keeping past presidents and vice presidents wealthy?," AOL, January 17, 2017; Memo from Stan Greenberg to the Gore/Lieberman Team, Greenberg Quinlan Research Inc., October 29, 2000; Miller Center,

"Interview with Roy Neel," University of Virginia, November 14, 2002; William Schneider, "Elián González Defeated Al Gore," *The Atlantic*, May 2, 2001; David Barstow and Katharine Q. Seelye, "The 2000 Campaign: The Selection; In Selecting a No. 2, No Detail Too Small," *New York Times*, August 9, 2000; Email from Lowell A. Weiss to Al Gore June 3, 1999, Bill Clinton Presidential Library, White House communication concerning Vice President Gore's campaign for office; Alexander C. Kaufman, "Al Gore's Stupendous Wealth Complicated his Climate Change Message. That Can Change," *Huffington Post*, August 18, 2017; John Judis, "Al Gore and the Temple of Doom," *The American Prospect,* December 19, 2001; Email from Pauline LaFon Gore to her son, Al Gore, Bill Clinton Presidential Library, White House communication concerning Vice President Gore's campaign for office.

XI. The Shadow President: Cheney and His Sidekick Bush

Interview subjects include Dick Cheney, Laura Bush, Neil Patel, Chris Hill, Donald Rumsfeld, Lucy Tutwiler, Richard Moe, and Tony Fratto. Published material includes: Dick Cheney with Liz Cheney, *In My Time: A Personal and Political Memoir* (New York: Simon & Schuster, 2011); "Bush, Cheney meet with 9/11 panel," CNN Politics, April 30, 2004, http://www.cnn.com/2004/ALLPOLITICS/04/29/bush.911.commission/; *Frontline,* "The Secret History of ISIS," PBS; Ron Suskind, *The One Percent Doctrine: Deep Inside America's Pursuit of its Enemies Since 9/11* (New York: Simon & Schuster, 2006); Peter Baker, *Days of Fire: Bush and Cheney in the White House* (New York: Random House, 2013); Collections at the Dole Archives, University of Kansas, http://dolearchivecollections .ku.edu/collections/vip_letters/c020_010_000_057.pdf; Rebecca Savransky, "Dick Cheney has yet to apologize to the man he shot in the face," *The Hill,* February 11, 2016; James Mann, "The Armageddon Plan," *The Atlantic*, March 2004; Dana Bash, "Cheney accidentally shoots fellow hunter," CNN, February 13, 2006; Sanjay Gupta, "Dick Cheney's Heart," *60 Minutes*, CBS, October 20, 2013; James Rosen, *Cheney One on One: A Candid Conversation with America's Most Controversial Statesman* (New York: Regnery, 2015); David Makovsky, "The Silent Strike: How Israel bombed a Syrian nuclear installation and kept it secret," *The New Yorker*, September 17, 2012; Mark Updegrove, *The Last Republicans: Inside the Extraordinary Relationship Between George H. W. Bush and George W. Bush* (New York: Harper, 2017); Barton Gellman and Jo Becker, "A Different Understanding with the President," *Washington Post*, June 24, 2007; Carol D. Leoning,

Shane Harris, and Greg Jaffe, "Breaking with tradition, Trump skips President's written intelligence report and relies on oral briefings," *Washington Post*, February 9, 2018.

XII. Fanboy: The Love Story of Joe and Barack

Interview subjects include Joe Biden, Valerie Jarrett, David Axelrod, Josh Earnest, Tony Blinken, Steve Ricchetti, Shailagh Murray, Ted Kaufman, Jay Carney, Alyssa Mastromonaco, Bob Dole, Ron Klain, Chris Hill, Sarah Bianchi, Bruce Reed, Mike Donilon, Adam Frankel, Dylan Loewe, and Tony Fratto. Published material includes: Evan Osnos, "The Biden Agenda: Reckoning with Ukraine and Iraq, and keeping an eye on 2016," *The New Yorker*, July 28, 2014; Joe Biden, *Promises to Keep* (New York: Random House, 2007); Jonathan Alter, "I Wish to Hell I'd Just Kept Saying the Exact Same Thing," *New York Times Magazine*, January 17, 2017; Maureen Dowd, "Joe Biden in 2016: What Would Beau Do?," *New York Times*, August 1, 2015; Joe Biden, *Promise Me, Dad: A Year of Hope, Hardship, and Purpose* (New York: Flatiron Books, 2017); Jonathan Allen and Amie Parnes, *Shattered: Inside Hillary Clinton's Doomed Campaign* (New York: Crown, 2017); Nicholas Fandos, "Joe Biden's Role in '90s Crime Law Could Haunt Any Presidential Bid," *New York Times*, August 21, 2015; Samantha Cooney, "Joe Biden Says He Owes Anita Hill an Apology," *Time*, December 14, 2017; Jordan Fabian and Amie Parnes, "Biden contradicts Clinton's account of bin Laden raid decision," *The Hill*, October 20, 2015; Glenn Thrush, "Clinton Revamps Stump Speech to Tout Her Role in Bin Laden Raid," *Politico*, January 1, 2016; John M. Broder, "Biden's Record on Race Is Scuffed by 3 Episodes," *New York Times*, September 17, 2008; C-SPAN, "Evolution of the Vice Presidency, October 20, 2015, https://www.c-span.org/video/?328827–1/walter-mondale-joe-biden-reflections-office-vice-president; The Situation Room Transcript, November 13, 2008, CNN, http://www.cnn.com/TRANSCRIPTS/0811/13/sitroom.02.html; Ed O'Keefe and Paul Kane, "The week John McCain shook the Senate," *Washington Post*, July 28, 2017; Roxanne Roberts, "Joe Biden still wants to be president, can his family endure one last campaign?," *Washington Post*, July 30, 2017; Mark Landler, "The Afghan War and the Evolution of Obama," *New York Times*, January 1, 2017; Vice President Biden Discusses Grief at TAPS, May 25, 2012, https://www.youtube.com/watch?v=GwZ6UfXm410; Josh Gerstein, "Was Biden Right?," *Politico*, June 6, 2013; Katherine Skiba, "Sasha Obama, Biden's granddaughter forge friendship," *Chicago Tribune*, July 18, 2015; Joe Biden, "We Are Living Through a Battle for the Soul of This Nation," *The Atlantic*, August 27, 2017;

Hillary Clinton, *What Happened* (New York: Simon & Schuster, 2017); David Rutz, "Biden Again Points Out Military Aide Carrying Nuclear Codes at Clinton Rally," *Free Beacon*, September 1, 2016; Amy Davidson Sorkin, "Why Joe Biden Should Run," *The New Yorker*, June 12, 2015; Jeanne Marie Laskas, "Have You Heard the One About President Joe Biden?" *GQ*, July 18, 2013; Alexander Marquardt, "Biden predicts early crisis will test Obama," CNN Political Tracker, October 20, 2008; Richard Socarides, "Forcing Obama's Hand on Gay Marriage," *The New Yorker*, April 15, 2014; Philip Rucker, "Donna Brazile: I considered replacing Clinton with Biden as 2016 Democratic Nominee," *Washington Post*, November 4, 2017.

XIII. Man on a Wire: Mike Pence's Tightrope Act

Interview subjects include Greg Pence, Nick Ayers, Dan Coats, Josh Pitcock, Marc Lotter, Dan Quayle, A. B. Culvahouse, Bill Smith, Bill Oesterle, Jim Atterholt, Ed Simcox, Joe Biden, Representative Marsha Blackburn, Greg Garrison, Gary Varvel, Jeff Cardwell, Kent Sterling, Steve Simpson, Phil Sharp, Bill Kristol, Senator Roy Blunt, David McIntosh, Brendan Buck, and Christy Denault. Published material includes: Darren Samuelsohn, "The old cassettes that explain Mike Pence," *Politico*, July 20, 2016; Harry McCawley, "The Mike Pence story: From a youth in Columbus to candidate for vice president," *The Republic*, July 17, 2016; Noah Bierman, "Vice President Mike Pence stays loyal to Trump, but it could come at a cost," *Los Angeles Times*, July 3, 2017; Jonathan Mahler and Dirk Johnson, "Mike Pence's Journey: Catholic Democrat to Evangelical Republican," *New York Times,* July 20, 2016; Maureen Groppe and Tony Cook, "How Vice President Pence is weathering the Donald Trump Jr. Russia revelations," *Indianapolis Star*, July 12, 2017; McKay Coppins, "God's Plan for Mike Pence: Will the vice president—and the religious right—be rewarded for their embrace of Donald Trump?" *The Atlantic*, January/February 2018; "Karen Pence's 1991 letter to the editor," *Washington Post*, March 28, 2017; Shari Rudavsky, "Karen Pence is right at home," *Indianapolis Star*, December 12, 2013; Derek Hawkins, "Trump reportedly mocked Pence, joked about him wanting to 'hang' gays," *Washington Post*, October 17, 2017; Melissa Langsam Braunstein, "Second Lady Karen Pence Opens Up About Her Struggles with Infertility," thefederalist.com; Philip Elliot, "The Persistent Passion of Vice President Mike Pence," *Time,* May 25, 217; Stephen Rodrick, "The Radical Crusade of Mike Pence," *Rolling Stone*, January 23, 2017; Maggie Severns, Matthew Nussbaum, and Ben LeFebvre, "Pence groups rakes in corporate PAC money," *Politico*, September 26, 2017; Craig Fehrman, "Where Is Mike Pence's Faith?," *Slate*, July 16,

2016; Lesley Stahl, "The Republican Ticket: Trump and Pence," *60 Minutes*, CBS, July 17, 2016, https://www.cbsnews.com/news/60-minutes-trump -pence-republican-ticket/; Jane Mayer, "The Danger of President Pence," *The New Yorker,* October 23, 2017; Rep. Mike Pence, Indiana, Contributors 1989–2014, https://www.opensecrets.org/members-of-congress/contributors ?cycle=Career&type=I&cid=N00003765&newMem=N&recs=20; Carol E. Lee, Kristen Welker, Stephanie Ruhle, and Dafna Linzer, "Tillerson's Fury at Trump Required an Intervention from Pence," October 4, 2017, https://www.nbcnews.com/politics/white-house/tillerson-s-fury-trump -required-intervention-pence-n806451; Chelsea Schneider and Tony Cook, "Pence signs new abortion restrictions into law with a prayer," *Indianapolis Star,* March 25, 2016; Jonathan Martin and Alexander Burns, "Republican Shadow Campaign for 2020 Takes Shape as Trump Doubts Grow," *New York Times,* August 6, 2017.

Epilogue: A Heartbeat Away

Interviews include Joe Biden, Al Gore, Tipper Gore, Dan Quayle, Dick Cheney, Valerie Jarrett, Bill Kristol, and Greg Zoeller.

Photo Insert Sources and Credits

INSERT ONE: National Park Service photo, courtesy Eisenhower Presidential Library & Museum; National Park Service photo, courtesy Eisenhower Presidential Library & Museum; courtesy John F. Kennedy Presidential Library and Museum, Boston; Robert Knudsen/White House, courtesy John F. Kennedy Presidential Library and Museum, Boston; City News Bureau, courtesy John F. Kennedy Presidential Library and Museum, Boston; courtesy Lyndon B. Johnson Library; Robert Knudsen/White House, courtesy Richard Nixon Presidential Library; William Parish/White House, courtesy Richard Nixon Presidential Library; Oliver Atkins/ White House, courtesy, Richard Nixon Presidential Library; courtesy Gerald R. Ford Presidential Library; David Hume Kennerly/White House, courtesy Gerald R. Ford Presidential Library; David Hume Kennerly/ White House, courtesy Gerald R. Ford Presidential Library; Robert Mc-Neely, courtesy Minnesota History Center; Getty Images/Mark Wilson; courtesy George Bush Presidential Library and Museum; courtesy George Bush Presidential Library and Museum; courtesy George Bush Presidential Library and Museum; Getty Images.

INSERT TWO: Associated Press/Stephan Savoia; Associated Press/Doug Mills; William Vasta, courtesy National Archives and Records Administration;

Presidential Materials Division, National Archives and Records Administration/ Molly Bingham; courtesy National Archives and Records Administration; courtesy National Archives and Records Administration; courtesy National Archives and Records Administration, Collection: Vice Presidential Records of the Photography Office (George W. Bush Administration), 1/20/2001–1/20/2009; courtesy National Archives and Records Administration, Collection: Vice Presidential Records of the Photography Office (George W. Bush Administration), 1/20/2001–1/20/2009; David Bohrer/ White House, courtesy National Archives and Records Administration; Associated Press/Morry Gash; Getty Images/Charles Ommanney; Pete Souza/ White House, Barack Obama Presidential Library; Pete Souza/White House, Barack Obama Presidential Library; Pete Souza/White House, Barack Obama Presidential Library; Pete Souza/White House, Barack Obama Presidential Library; Pete Souza/White House, Barack Obama Presidential Library; Lawrence Jackson/White House, Barack Obama Presidential Library; Pete Souza/White House, Barack Obama Presidential Library; Associated Press/ Mary Altaffer; Getty Images/Mark Wilson; Bao N. Huynh/White House; Shealah Craighead/White House; Official White House Photo/Office of the Vice President; Myles Cullen/White House.

Agnew, Spiro. *Go Quietly . . . or else.* New York: William Morrow and Company, Inc., 1980.

———. *Where He Stands.* New York: Hawthorn Books, Inc., 1968.

Baime, A. J. *The Accidental President: Harry S. Truman and the Four Months that Changed the World.* New York: Houghton Mifflin Harcourt, 2017.

Baker, Peter. *Days of Fire: Bush and Cheney in the White House.* New York: Doubleday, 2013.

Bernstein, Carl. *A Woman in Charge: The Life of Hillary Rodham Clinton.* New York: Vintage Books, a division of Random House, Inc., 2007.

Biden, Joe. *Promise Me, Dad: A Year of Hope, Hardship, and Purpose.* New York: Flatiron Books, 2017.

———. *Promises to Keep: On Life and Politics.* New York: Random House, 2007.

Branch, Taylor. *The Clinton Tapes: Wrestling History with the President.* New York: Simon & Schuster, 2009.

Bush, Barbara. *A Memoir: Barbara Bush.* New York: Scribner, 1994.

Bush, Laura. *Spoken from the Heart.* New York: Simon & Schuster, 2010.

Cannon, James. *Gerald R. Ford: An Honorable Life.* Ann Arbor: University of Michigan Press, 2013.

Caro, Robert A. *The Years of Lyndon Johnson: Master of the Senate.* New York: Alfred A. Knopf, 2002.

———. *The Years of Lyndon Johnson: The Passage of Power.* New York: Vintage Books, 2012.

Cheney, Dick. *In My Time.* New York: Simon & Schuster, 2011.

Clinton, Hillary Rodham. *Living History.* New York: Scribner, 2003.

———. *What Happened.* New York: Simon & Schuster, 2017.

DeFrank, Thomas M. *Write It When I'm Gone: Remarkable Off-the-Record Conversations with Gerald R. Ford.* New York: G.P. Putnam's Sons, 2007.

Doyle, William. *Inside the Oval Office: The White House Tapes from FDR to Clinton.* New York: Kodansha International, 1999.

Ehrlichman, John. *Witness to Power: The Nixon Years.* New York: Simon & Schuster, 1982.

Eisenhower, Julie Nixon. *Pat Nixon: The Untold Story.* New York: Simon & Schuster, 2007.

Fields, Alonzo. *My 21 Years in the White House.* New York: Crest Books, 1961.

Ford, Betty, with Chris Chase. *The Times of My Life.* New York: Harper & Row Publishers and The Reader's Digest Association, Inc., 1978.

Ford, Gerald. *A Time to Heal.* New York: Harper & Row, 1979.

Frank, Jeffrey. *Ike and Dick: Portrait of a Strange Political Marriage.* New York: Simon & Schuster, 2013.

Gellman, Barton. *Angler: The Cheney Vice Presidency.* New York: Penguin, 2009.

Gibbs, Nancy, and Michael Duffy. *The Presidents Club.* New York: Simon & Schuster, 2012.

Gillette, Michael L. *Lady Bird Johnson: An Oral History.* New York: Oxford University Press, 2012.

Goldstein, Joel. *The White House Vice Presidency: The Path to Significance, Mondale to Biden.* Lawrence, KS: University Press of Kansas, 2016.

Goodwin, Doris Kearns. *Lyndon Johnson and the American Dream.* New York: St. Martin's Griffin, 1991.

Haldeman, H. R. *The Haldeman Diaries: Inside the Nixon White House.* New York: G.P. Putnam's Sons, 1994.

Hayes, Stephen F. *Cheney: The Untold Story of America's Most Powerful and Controversial Vice President.* New York: HarperCollins, 2007.

Hill, Clint, and Lisa McCubbin. *Five Days in November.* New York: Gallery Books, a division of Simon & Schuster, 2013.

——. *Mrs. Kennedy and Me.* New York: Gallery Books, a division of Simon & Schuster, 2012.

Isenberg, Nancy. *Fallen Founder: The Life of Aaron Burr.* New York: Viking, 2007.

Johnson, Lady Bird. *A White House Diary.* New York: Holt, Rinehart and Winston, 1970.

Jones, Charles O. *The American Presidency: A Very Short Introduction.* Oxford University Press, 2007.

Lincoln, Evelyn. *Kennedy and Johnson.* New York: Holt, Rinehart and Winston, 1968.

Manchester, William. *The Death of a President: November 20–November 25, 1963*. New York: Harper & Row, 1967.

Maraniss, David. *The Prince of Tennessee*. New York: Simon & Schuster, 2000.

Meacham, Jon. *Destiny and Power: The American Odyssey of George Herbert Walker Bush*. New York: Random House, 2015.

Oliphant, Thomas, and Curtis Wilkie. *The Road to Camelot: Inside JFK's Five-Year Campaign*. New York: Simon & Schuster, 2017.

Pearson, Drew. *Washington Merry-Go-Round: The Drew Pearson Diaries, 1960–1969*. Lincoln, NE: Potomac Books, University of Nebraska Press, 2015.

Quayle, Dan. *Standing Firm: A Vice-Presidential Memoir*. New York: Harper-Collins, 1994.

Reagan, Nancy, with William Novak. *My Turn: The Memoirs of Nancy Reagan*. New York: Random House, 1989.

Reagan, Ronald. *The Reagan Diaries*. New York: HarperCollins, 2007.

Regan, Donald T. *For the Record*. New York: Harcourt Brace Jovanovich, Publishers, 1988.

Rice, Condoleezza. *No Higher Honor: A Memoir of My Years in Washington*. New York: Random House, 2011.

Rosen, James. *Cheney One on One: A Candid Conversation with America's Most Controversial Statesman*. Washington, D.C.: Regnery Publishing, 2015.

Schlesinger, Arthur M., Jr. *Jacqueline Kennedy: Historic Conversations on Life with John F. Kennedy*. New York: Hyperion, 2011.

Suskind, Ron. *The One Percent Doctrine: Deep Inside America's Pursuit of Its Enemies Since 9/11*. New York: Simon & Schuster, 2006.

Truman, Harry. *1945: Year of Decisions: Memoirs, Volume 1*. New York: New Word City, 2014.

Updegrove, Mark. *The Last Republicans: Inside the Extraordinary Relationship Between George H. W. Bush and George W. Bush*. New York: Harper, 2017.

Weidenfeld, Sheila Rabb. *First Lady's Lady: With the Fords at the White House*. New York: G.P. Putnam's Sons, 1979.

Witcover, Jules. *The American Vice Presidency: From Irrelevance to Power*. Washington, D.C.: Smithsonian Books, 2014.

———. *Joe Biden: A Life of Trial and Redemption*. New York: William Morrow and Company, Inc., 2010.

———. *Very Strange Bedfellows: The Short and Unhappy Marriage of Richard Nixon and Spiro Agnew*. New York: Public Affairs, 2007.

Woodward, Bob. *The Last of the President's Men*. New York: Simon & Schuster, 2015.

Woodward, Bob, and Carl Bernstein. *The Final Days*. New York: Simon & Schuster, 1976.

Woodward, Bob, and David S. Broder, *The Man Who Would Be President*. New York: Simon & Schuster, 1992.

INDEX

Domenici, Pete, 39
Domestic Policy Council, 163
Donilon, Mike, 7, 224, 232, 237, 251–52
Dowd, Maureen, 226
Dukakis, Michael, 186

Eagleton, Thomas, 43–44
Early, Steve, 104–5, 251
Earnest, Josh, 16, 227
Edwards, John, 49, 98, 207, 221
Ehrlichman, John, 153, 159
Eisenhower, Dwight D., 15, 115, 129–
    39, 150, 161, 176, 206–7, 233, 292
Eisenhower Executive Office Build-
    ing, 10, 15, 81, 115, 127, 244, 278
elections
    of 1796, 19
    of 1800, 19
    of 1824, 184
    of 1836, 17, 173
    of 1876, 184
    of 1888, 184
    of 1940, 20
    of 1944, 106
    of 1948, 131
    of 1950, 131
    of 1952, 133–33
    of 1954, 135
    of 1956, 135
    of 1960, 17, 25–28, 32–33, 137,
        140–41, 197
    of 1964, 21, 117, 145, 149, 162
    of 1968, 17, 70, 137, 146–47, 149–
        50, 162
    of 1972, 43–44, 155, 157, 162, 230,
        233
    of 1974, 44, 155
    of 1976, 34, 44, 157, 164–65, 169–
        70, 171

of 1980, 33–34, 44, 172, 271
of 1984, 42–43, 172
of 1988, 17, 45, 51, 177, 186, 227, 272
of 1990, 272
of 1992, 45–46, 93, 178–81, 186–
    87, 190–91, 193
of 1996, 198
of 2000, 17, 21, 47–48, 92–93, 95–
    96, 98, 183–86, 189–90, 193–94,
    196, 198–202, 204, 275, 295
of 2004, 48–49, 96, 98–99, 204,
    207, 212, 221, 235, 281
of 2006, 276
of 2008, 7, 49–50, 206, 234–39,
    242, 293–94
of 2012, 59, 61, 228–29, 243, 248,
    250–52, 275, 281
of 2016, 59–68, 183–84, 224–32,
    247, 275, 283, 285–87, 289
of 2018, 285
of 2020, 225, 232, 266, 285
Electoral College, 19, 183–85
Elizondo, Carlos, 76, 101–2
Elmendorf, Steve, 230
El Salvador, 258
Emanuel, Rahm, 51, 206
Enola Gay (B-29 bomber), 107
Environmental Protection Agency
    (EPA), 180
Ernst, Joni, 59
Ernst, Max, 72
Eskew, Carter, 184–85, 189–90, 199–
    200
evangelical Christianity, 83, 178, 272,
    278
Executive Office Building, 142
Exide Corp., 99
ExxonMobil, 269

ABOUT THE AUTHOR

Kate Andersen Brower is the author of the number one *New York Times* bestseller *The Residence* and the *New York Times* bestseller *First Women*. After working for CBS News in New York and Fox News in Washington, D.C., she covered the White House for Bloomberg News during Barack Obama's administration. Now a CNN contributor, Brower has written for the *New York Times*, *Vanity Fair*, *Time*, the *Washington Post*, and *Bloomberg Businessweek*. She lives outside Washington, D.C., with her husband, their two young children, and their wheaten terrier.